Flow control of thermal convection using thermo el

forces in a cylindrical annulus

Flow control of thermal convection using thermo electro hydrodynamic forces in a cylindrical annulus

Von der Fakultät für Maschinenbau, Elektro- und Energiesysteme der Brandenburgischen Technischen Universität Cottbus–Senftenberg zur Erlangung des akademischen Grades eines Dr.-Ing.

genehmigte Dissertation

vorgelegt von

M.Sc.
Marcel Jongmanns
geboren am 31.01.1985 in Nettetal

Vorsitzende: Sabine Weiß
Gutachter: Christoph Egbers
Gutachter: Innocent Mutabazi
Tag der mündlichen Prüfung: 11.04.2019

Bibliografische Information der Deutschen Nationalbibliothek
Die Deutsche Nationalbibliothek verzeichnet diese Publikation in der Deutschen Nationalbibliografie; detaillierte bibliografische Daten sind im Internet über http://dnb.d-nb.de abrufbar.

1. Aufl. - Göttingen: Cuvillier, 2019
Zugl.: (BTU) Cottbus-Senftenberg, Univ., Diss., 2019

Nonnenstieg 8, 37075 Göttingen
Telefon: 0551-54724-0
Telefax: 0551-54724-21
www.cuvillier.de

1. Auflage, 2019
Gedruckt auf umweltfreundlichem, säurefreiem Papier aus nachhaltiger Forstwirtschaft

ISBN 978-3-7369-7044-1
eISBN 978-3-7369-6044-2

Contents

List of Figures

List of Tables

Used abbreviations and symbols

Abbreviations

CNES	Centre national d'études spatiales French aerospace organization
DEP	Dielectrophoretic
DLR	Deutsches Zentrum für Luft- und Raumfahrt German aerospace organization
ESA	European Space Agency
ESDP	Experiment Safety Data Package
LLS	Laser Light Sheet
PFC	Parabolic Flight Campaign
PIV	Particle Image Velocimetry
PMMA	Polymethylmethacrylate
TEHD	Thermo electro hydrodynamic

Mathematical and physical symbols

symbol	unit	description
α	$\frac{1}{K}$	Thermal expansion coefficient
α_e	$\frac{1}{K}$	Thermal expansion coefficient for TEHD
ε	$\frac{As}{Vm}$	Electric permittivity, dielectric constant
ε_r		Relative electric permittivity, relative dielectric constant
Γ		Aspect ratio
η		Radius ratio
ΔT	K	Temperature difference
κ	$\frac{m^2}{s}$	Thermal diffusivity
λ	$\frac{W}{mK}$	Thermal conductivity
μ	Pas	Dynamic viscosity
ν	$\frac{m^2}{s}$	Kinematic viscosity
ρ	$\frac{kg}{m^3}$	Density
ρ_e	$\frac{C}{m^3}$	Free electric charge density
σ	$\frac{S}{m}$	Electric conductivity
τ_e	s	Charge relaxation time
τ_k	s	Thermal relaxation time
τ_v	s	Viscous relaxation time

Θ	K	Very small temperature variation
A	m^2	Area
d	m	Distance, gap width
E	$\frac{V}{m}$	Electric field
F	N	Force
F_C	N	Coulomb (electrophoretic) force
F_D	N	Dielectric force
F_{DEP}	N	Dielectrophoretic force
F_{ES}	N	Electrostrictive force
g_e	$\frac{m}{s^2}$	Electric gravity
H	m	Height
p	Pa	Pressure
P	W	power
$\dot{q}_{cond}$	W	Conductive heat transfer
$\dot{q}_{conv}$	W	Convective heat transfer
r	m	Radius
R_1	m	Radius of inner cylinder
R_2	m	Radius of outer cylinder
t	s	Time
T	K	Temperature
V_{peak}	V	Peak Voltage
V_{rms}	V	Root-mean-square Voltage
$\mathcal{G}r$		Grashof number
$\mathcal{N}u$		Nusselt number
$\mathcal{P}r$		Prandtl number
$\mathcal{R}a$		Rayleigh number
$\mathcal{R}a_e$		electric Rayleigh number

symbol	value	description
ε_0	$8.85 * 10^{-12} \frac{As}{Vm}$	Electric permittivity in vacuum
e	$10^{-3} \ldots 10^{-2} \frac{1}{K}$	Thermal coefficient of permittivity [1]
g	$9.81 \frac{m}{s^2}$	Earth's gravitational acceleration

[1] Estimate for AK5 (Yoshikawa *et al.* 2013)

Zusammenfassung

Innerhalb dieser Arbeit wird der Einfluss der dielektrophoretischen (DEP) Kraft auf eine thermische Strömung in einem zylindrischen Annulus betrachtet.

Für die Durchführung der Experimente wurde ein spezieller Aufbau entworfen und umgesetzt. Da der Aufbau in Parabelflügen eingesetzt wird müssen höhere Sicherheitsstandards eingehalten werden als in Laborversuchen. Es wurden zwei Experimentboxen für unterschiedliche Messmethoden entworfen. Innerhalb der Boxen befinden sich unter anderem die eigentlichen Experimentzellen, die aus zwei konzentrischen Zylindern bestehen. Es wurden Zellen mit unterschiedlichen Höhen, 30mm und 100mm, betrachtet. Der innere Zylinder wird geheizt und ist mit der Phase eines Hochspannungsverstärkers verbunden. Der äußere Zylinder wird gekühlt und ist mit Masse verbunden. Der Spalt ist mit Silikonöl AK5 gefüllt, welches als Dielektrikum dient.

Der Temperaturunterschied führt zu einer thermischen Konvektion in axialer Richtung im Fluid. Die DEP Kraft führt zu einer radial nach innen gerichteten Kraftwirkung. Die Kombination beider Kräfte führt zu einer komplexen Strömungsstruktur in Laborversuchen. Daher wurden die Experimente auch unter Schwerelosigkeit in Parabelfügen durchgeführt, wodurch die axiale Kraftwirkung verschwindet. Die Strömungsmuster die unter Schwerelosigkeit zu sehen sind können von den Mustern die unter Laborbedingungen zu sehen sind abweichen, obwohl die anderen Experimentparameter gleich sind. Die Visualisierung erfolgt mit der Shadowgraph Methode, Synthetic Schlieren Methode oder PIV. Zusätzlich werden Temperaturmessungen durchgeführt um den konvektiven Wärmetransport über die Nusselt Zahl $\mathcal{N}u$ zu bestimmen. Die Ergebnisse der Experimente werden mit den Ergebnissen einer linearen Stabilitätsanalyse verglichen. Die durch die Theorie vorherbestimmten Strömungsmuster konnten durch die Experimente nachgewiesen werden. Es ist ebenfalls ein Zusammenhang zwischen dem Strömungsmuster und $\mathcal{N}u$ zu sehen.

Summary

The main topic of this thesis is the influence of the dielectrophoretic (DEP) force on a thermal flow in a cylindrical annulus.

To perform these experiments an experimental setup was designed and constructed. Since it was used in parabolic flights certain safety standards and limitations had to be applied. Two different Experiment Boxes were build for different measurement methods. Inside these boxes are among other things the Experiment Cells, which consist of two concentric cylinders with a height of $H = 30mm$ or $H = 100mm$. The inner cylinder is heated and connected to the phase of a high-voltage amplifier. The outer cylinder is cooled and connected to ground. The gap is filled with silicone oil AK5, which is a dielectric fluid.

The temperature difference causes a thermal convection in axial direction. The DEP force invoked by the electric field creates a radial, inward-directed force. The combination of both forces creates a complex flow pattern in laboratory experiments. The experiment is also performed in a microgravity environment during parabolic flight campaigns to reduce the axial force to $< 0.1g$. The flow patterns which emerge during microgravity conditions can be different than in laboratory experiments with the same applied temperature gradient and voltage. These patterns are visualized using different methods such as Shadowgraph, Synthetic Schlieren and PIV. Additionally the convective heat transfer is determined and described using the Nusselt number $\mathcal{N}u$. The experimental data is compared to theoretical data from a linear stability analysis. The patterns predicted by the theory can also be found in the experiments. It is also possible to see a change in $\mathcal{N}u$ when the flow pattern changes.

1. Introduction

Electric fields created by direct current (d.c.) have been used in microfluidics and biology to e.g. sort cells or purify fluids by electrophoresis since a long time. However, the application of alternating current (a.c.) fields has not been researched as thoroughly. In this thesis I will present some results of experiments on the effect of an a.c. electric field on a thermal convection in a dielectric fluid.

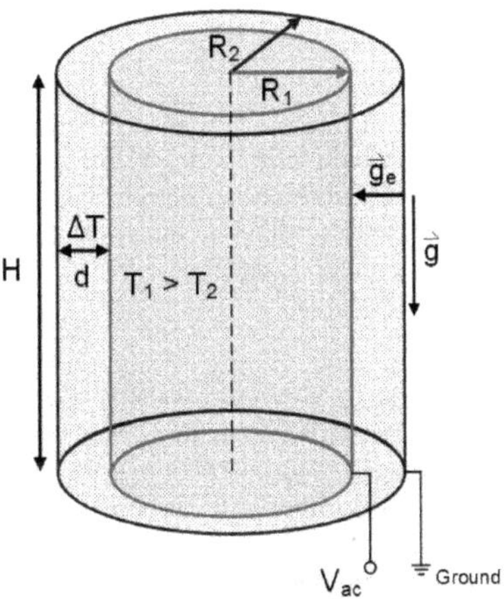

Figure 1.1.: A schematic of the experimental setup.

The experiments are performed in a cylindrical annulus (Fig. 1.1). The inner cylinder is heated, connected to the phase of a high-voltage amplifier and has the radius R_1. The outer cylinder is cooled, connected to electric ground and has the radius R_2. The gap with width $d = R_2 - R_1$ is filled with a dielectric fluid. The height of the cylinders are given by H. When a temperature difference $\Delta T = T_1 - T_2$ is applied, then a flow with an unicellular flow pattern (Fig. 1.2) is induced. This flow is driven by the temperature dependent density changes of the fluid and the natural gravity g. The permittivity of the fluid is also temperature dependent. It increases as the temperature of the fluid decreases. When an electric a.c. field is applied in addition to the temperature, then an electric gravity g_e acts on the force. It is in fact not a gravity like earth's gravity, but the effects can be compared. The permittivity increases when the temperature decreases. The fluid from the outer cylinder, which is cooled, will be pulled towards the inner cylinder (Pohl 1978). This is analogous to a density gradient and the gravity of the earth. Since the experiment is performed in a cylindrical annulus, the electric field is non-uniform and stronger near the inner cylinder. The combination of thermal and electric forces results in the thermo electrohydrodynamic force (TEHD). The experiment which we designed and build is able to achieve ΔT up to $20K$ and a peak voltage V_p up to $10kV$ at a frequency of up to $f = 500Hz$.

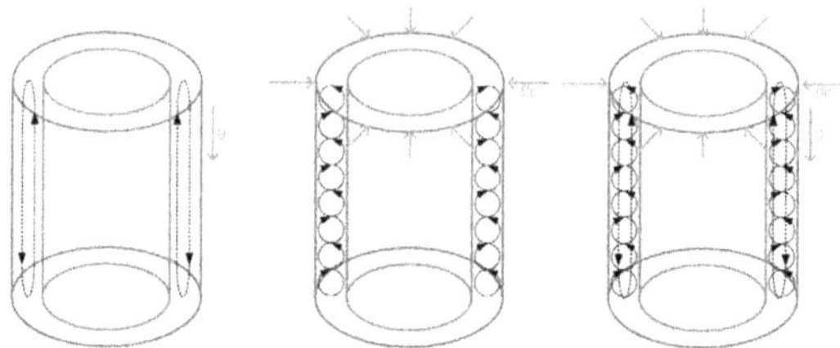

Figure 1.2.: The basic flow states inside the gap under different gravitational conditions (Dahley 2014). Left: Thermal convection with natural gravity. Middle: Thermal convection with applied artificial electric gravity g_e. Right: Superposition of both gravitational forces.

This flow is always superimposed by the natural convection in the laboratory. The natural gravity is stronger than the artificial electric gravity g_e (Eq. 3.5). Due to the curvature of the geometry the strength of g_e differs depending on the position in the gap (Fig. 1.4). It can be stronger than g directly at the inner cylinder, but is on average weaker. To isolate the effect of the radial oriented g_e, the experiments are performed in microgravity conditions during parabolic flights. Parabolic flights offer a comparably easy way to perform experiments in microgravity (μg). The experiment is mounted inside an aircraft and the pilots perform a special parabolic maneuver, which gives a phase of about $22s$ in μg ($g_z < 0.01g$). This time is very short compared to the time required that a stable temperature can establish, but it is enough time that new flow pattern can develop and stabilize (see Tbl. A.6).

This geometry was chosen because it resembles a triple pipe heat exchanger (Fig. 1.3). Hot fluid is pumped through the inner pipe. The cooling fluid is pumped through the outer pipe. The newly introduced middle pipe would contain the dielectric fluid. The heat transfer between the inner and outer pipe can then be regulated by changing the applied voltage. The experiments are performed in μg to get an understanding of the effect. Later this knowledge can be applied to design a setup to use in a $1g$ environment.

The experimental results are compared to a linear stability analysis (LSA)(Yoshikawa *et al.* 2013) and numerical simulations by Travnikov (Travnikov *et al.* 2015 and Olivier Crumeyrolle (not yet published). The LSA gives critical parameters for ΔT and V_p at which the flow changes the pattern in μg . There is also a LSA based on the same

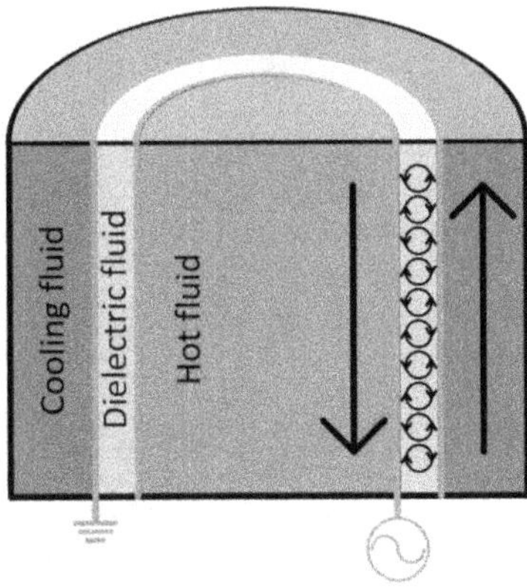

Figure 1.3.: A triple pipe heat exchanger where the TEHD force is used to control the convective heat transfer.

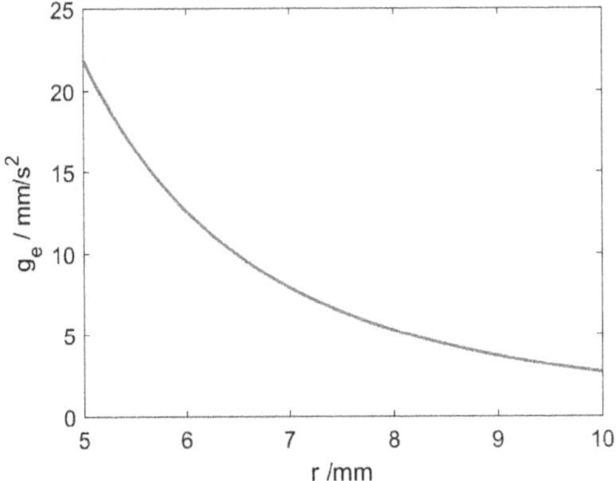

Figure 1.4.: The strength of the electric field inside the gap. The inner cylinder is at $r = 5mm$ and the outer at $r = 10mm$. The used parameters are $\Delta T = 10K$ and $V_p = 10kV$.

algorithm performed by Meyer, which gives the critical values for laboratory experiments including $g_z = 1g$ (Meyer 2017). The LSA assumes a cell of infinite length and applies the Boussinesq approximations for the density and permittivity. The simulations give a prediction how the 3D flow structure should look like in μg .

This effect gives a way to control a thermal convection. A gravitational force is required to create thermal convections. A possible application for TEHD is to create and control heat transfer in fluids in microgravity conditions. The gained knowledge can then be used to increase the efficiency of heat exchangers on earth (Bergles 1998; Futterer, Dahley, *et al.* 2016; Jongmanns, Meyer, *et al.* 2018; Laohalertdechaa *et al.* 2007; Marucho *et al.* 2013).

A first pre-study was performed by Norman Dahley (Dahley 2014) and Birgit Futterer (Futterer, Dahley, *et al.* 2016) at this department. I was able to use their experiences to continue to work on this topic. During the current project, we drastically improved the experimental setup to eliminate flashovers, modularized the setup to make it more variable, increase the reliability of the temperature gradients and measurements, and increased the quality of the visualization techniques. This allowed me to produces repeatable results with a higher quality, which are able to answer some of the open questions.

2. State of the art

Research on the DEP force has been performed by several groups since a long time, but it was mostly neglected in favor of the coulomb force. The coulomb force has applications in e.g. biology, where it is used to separate cells or molecules based on their charge. One of the first experiments involving the DEP force was done by Chandra and Smylie (Chandra *et al.* 1972). They used two concentric cylinders where the gap is filled with a silicone oil, one side is cooled and the other heated. An alternating electric field is applied to the two cylinders. The strength of the resulting DEP force is given as a dimensionless number, the electrical Rayleigh number. The temperatures were measured to calculate the Nusselt number, which is a measurement for convective heat transfer (Fig. 2.1). They also performed a linear stability analysis (LSA) to determine the critical parameters at which the heat transfer increases.

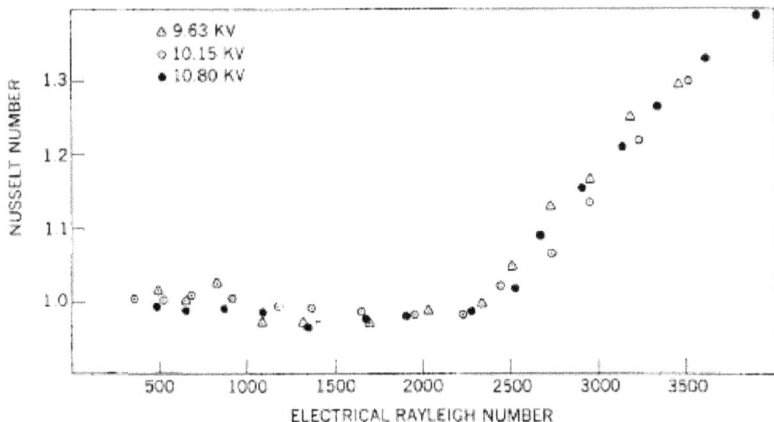

Figure 2.1.: Early experiment with applied electric field. The graph shows the critical Ra_e at which the convection, given by Nu, increases (Chandra *et al.* 1972).

The experimental results show that starting from an electric Rayleigh number of 2200 ± 100 the Nusselt number increases, which means that the heat transfer by convection increases because of the flow induced by the electric field. This agrees very well with their LSA, where the critical Rayleigh number is 2119.346. In their results, Nu first decreases slightly before is starts to increase. There is no explanation for this local minimum. The usage of DEP to increase heat transfer in fluid systems has also been suggested by others like Jones 1978, Laohalertdechaa *et al.* 2007, Futterer, Dahley, *et al.* 2016, and Allen *et al.* 1995.

A stability analysis (Bahloul *et al.* 2000) performed in a similar geometry, but without electrical field, aimed to determine the flow pattern within the gap along the height. The thermal buoyancy is characterized by the Grashof number, the diffusion properties of the fluid are defined by the Prandtl number and the geometry assumes a cylinder of infinite length with variable radius ratio η. They differentiate between two modes, a hydrodynamic (H.M.) and thermal mode (T.M.). The critical Grashof numbers for Prandtl numbers from 0 to 50 are given in Fig. 2.2. In the described thermal mode counter rotating vortices with their center at the center of the gap are established. This structure gradually changes to vortices with two rotation centers in the hydrodynamic mode.

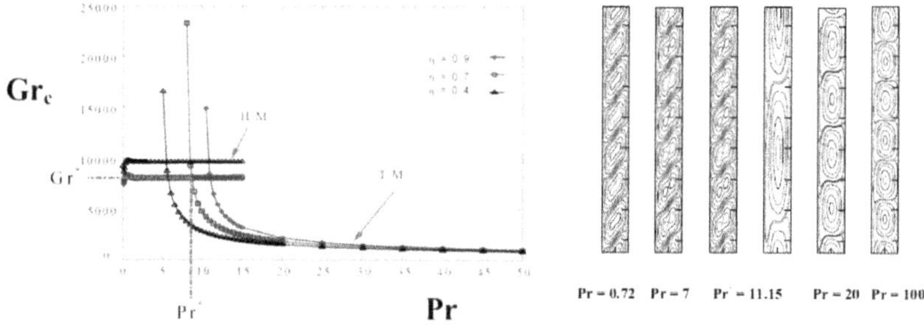

Figure 2.2.: Left: The point at Gr^* and Pr^* shows the parameters at which the mode changes between the hydrodynamic mode (H.M.) and thermal mode (T.M.) occurs. Right: Streamlines of the vortices (Bahloul *et al.* 2000).

The LSA is only concerned with the flow pattern, but not the resulting heat transfer. As different patterns have their specific effect on convective heat transfer, this information could indirectly be used to determine the heat transfer by convection. For example, a single, big convection cell has a lower convective heat transfer than several smaller cells.

Another approach looking at the effect of high voltage alternating electric fields on dielectric fluids was done by Sitte (Sitte 2004, Sitte and Rath 2003). Different methods were used. A first experiment was designed with a grounded electrode, which also emitted hot oil, and a plate to which the voltage was applied (Fig. 2.3, plate outside of image). The Shadowgraph method was used to visualize the jet. With increasing potential of the electric field the deflection angle of the jet increases, showing that the field has a direct influence on the fluid.

Some different parameters were tested by Takashima 1980. The temperature gradient was applied either by inward or outward heating and different radius ratios η were tested. For outward heating, the critical Ra_e decreased as η decreased. But for inward heating it was the other way around. The flow was more unstable for outward heating and can become completely stable for inward heating.

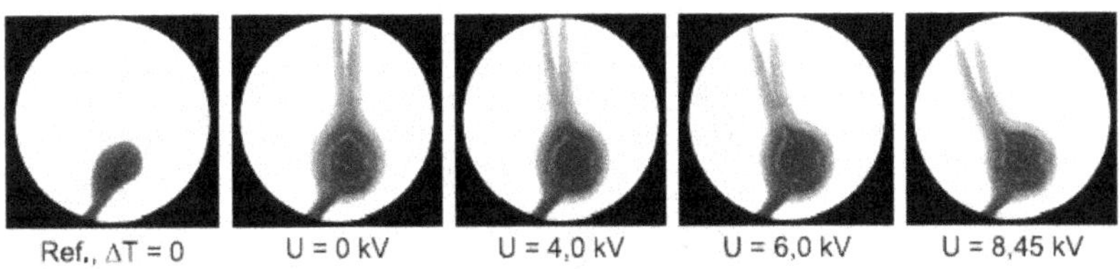

Figure 2.3.: Heated oil is emitted from an electrode. The angle of deflection increases with increasing electric potential (Sitte 2004).

Further Sitte did experiments in microgravity during a parabolic flight. For this experiments he used two concentric cylinders, similar to Chandra *et al.* 1972, where the gap is filled with a silicone oil. The gap is heated at one side and cooled on the other. The electric field is applied directly to the inner and outer cylinder. The Shadowgraph method is used as well to visualize the flow. There is no buoyancy flow in microgravity, since no external forces act on a fluid. However, when an electric a.c. field is applied to a fluid with stratified permittivity, a flow can be seen. The Shadowgraph images of Sitte give a first impression on how the structures inside the fluid changes (Fig. 2.4).

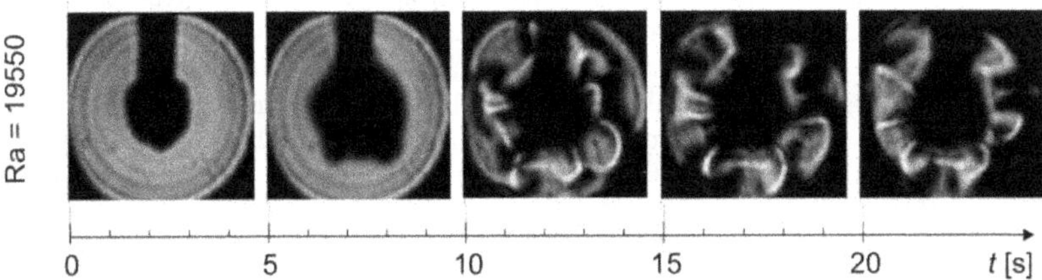

Figure 2.4.: The fluid inside the gap of two concentric cylinders is influenced with an electric field during a parabolic flight. The time line shows the development of the flow over one microgravity phase (Sitte 2004).

Quite recently a 3D numerical simulation (Travnikov *et al.* 2015) has been performed to compute the flow pattern in a cylindrical annulus of infinite length with applied temperature gradient and electric field. This simulation shows a pattern with slightly inclined columns from the bottom to the top of the gap (Fig. 2.5). The convective heat transfer was also determined and shows a greater increase than in a plane geometry.

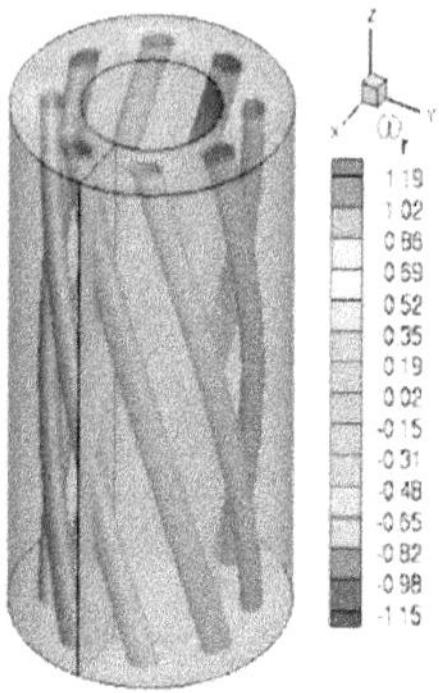

Figure 2.5.: A numerical simulation of the 3-dimensional flow assuming that the cell has infinite length with applied electric field and under microgravity conditions. $Ra_e = 1540, Pr = 100, \eta = 0.5$ (Travnikov *et al.* 2015).

A simulation of the development of the flow pattern in an experiment cell with $H = 30mm$, $\Gamma = 6$, $V_p = 10kV$ and $\Delta T = 10K$ has been performed by Olivier Crumeyrolle. He considered the same boundary conditions which are present during the parabolic flight. At the beginning of the μg phase (Fig. 2.6 left) there is still an unicellular convection pattern visible similar to the convection pattern in $1g$. At the middle of the μg phase this changes to a toroidal pattern, but also shows the onset of plumes in the lower part of the annulus. At the end of the μg phase, which is $22s$ after the μg phase started, there are only toroidal patterns at the upper and lower boundaries and plume structures otherwise.

Figure 2.6.: Flow patterns given by the isothermal surface over the course of one parabola of the parabolic flight (O. Crumeyrolle, internal communication). Left: Starting condition. Middle: After 10s in the μg phase. Right: At the end of the μg phase.

Martin Baumann and Philipp Gerstner also performed simulations of the flow pattern. Their main objective was the laboratory case (Seelig *et al.* 2019), but they also applied their code to the simulation to the μg case for an experiment cell with $H = 100mm$, $\Gamma = 20$, $V_p = 10kV$ and $\Delta T = 7K$ (data not published). Similar to Crumeyrolle they also considered the boundary conditions of the parabolic flight. The predicted flow pattern shows a toroidal pattern over the whole height of the gap at the end of the μg phase (Fig. 2.7)

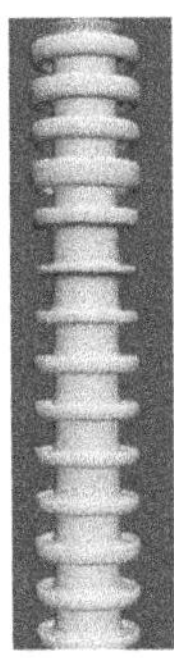

Figure 2.7.: Simulated flow pattern of an annulus with $H = 100mm$, $\Gamma = 20$, $V_p = 10kV$ and $\Delta T = 7K$ at the end of the μg phase. (P. Gerstner, internal communication).

A linear stability analysis was performed by Yoshikawa (Yoshikawa *et al.* 2013), which predicts the critical parameters at which the flow pattern changes. Results for the critical parameters exist for both, $1g$ and μg conditions. In a $1g$ environment two critical parameters are found (Fig. 2.8, Meyer, Jongmanns, *et al.* 2016). The first is at $\Delta T \approx$

$13.4K, V_{rms} \approx 2.1kV$, at which a columnar pattern is predicted. The second transition occurs at $\Delta T \approx 0.2K, V_{rms} \approx 9.2kV$, where the columns become inclined and a helical pattern is found. When the experimental parameters are below the critical parameters no perturbations should be visible and the flow pattern should be a unicellular base flow. In μg only the first set of critical parameters exist. When this is not reached, the external forces are too low and no convections are visible. The base state in μg is a conductive state, where the heat is transported by conduction only. When the critical parameters are surpassed the flow pattern is always expected to be helical. The critical parameters are dependent on the curvature of the system (Malik *et al.* 2012; Yoshikawa *et al.* 2013).

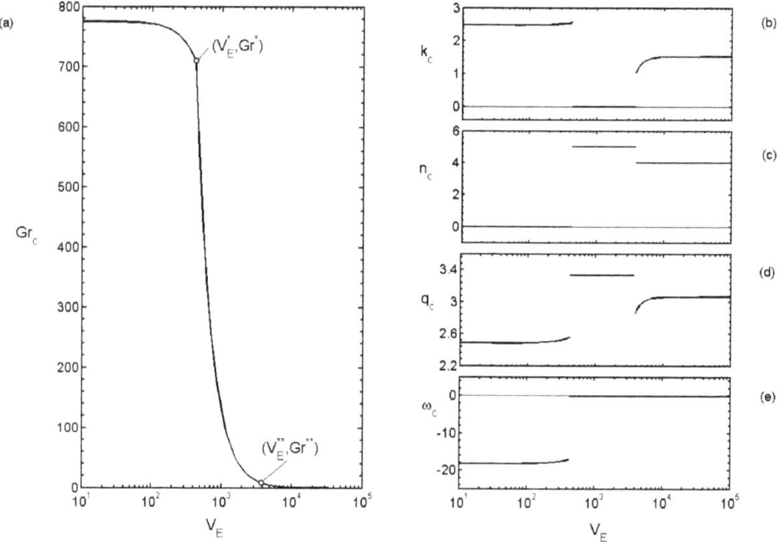

Figure 2.8.: Behavior of (a) the critical Grashof number, (b) the critical axial wavenumber, (c) the critical total wavenumber and (d) the critical frequency with the dimensionless electric potential for AK5 in a gap of $\eta = 0.5$. (Meyer, Jongmanns, *et al.* 2016).

Previous experiments were performed by Birgit Futterer and Norman Dahley (Futterer, Dahley, *et al.* 2016, Dahley 2014). A lot of the work was focused on the thermoelectrohydrodynamic effect in microgravity, but most published results are laboratory pre-experiments. Their work was focused on designing an experiment, which would allow to study the DEP effect in microgravity during parabolic flights. The Nu measurements in the laboratory is comparable to the data from Chandra and Smylie. As the voltage increases, Nu first decreases, but increases again at a critical voltage (Fig. 2.9).

This project, the examination of heat transfer and flow pattern in a cylindrical gap, started as side project of the GeoFlow experiment, but is now an independent project. The GeoFlow experiment applied a central force field to a spherical geometry using the DEP force (Futterer, Krebs, *et al.* 2013). The objective of this project is to simulate the environment of the Earth's core (GeoFlow I) and the Earth's mantle (GeoFlow II). It is fundamental research to get more understanding on how the flow inside the Earth behaves. The cylindrical experiment was designed with an actual application as heat enhancement effect in mind.

There are also numerical approaches to determine the flow pattern in a plate geometry (Fogaing *et al.* 2013). In this geometry, the electric field is homogeneous when the fluid is not heated. When a temperature gradient is applied the electric field becomes inhomogeneous due to the resulting permittivity gradient. However, the resulting DEP force is lower than in a cylindrical geometry, because there is no curvature. One of the premises for the DEP effect is to have an inhomogeneous. However, there is still an effect, but the required temperature gradients are a lot higher at $\Delta T > 40K$. The determined enhancement of the heat transfer for fluids with Pr $\geqslant 1$ is comparable to earlier experiments by Stiles *et al.* 1993.

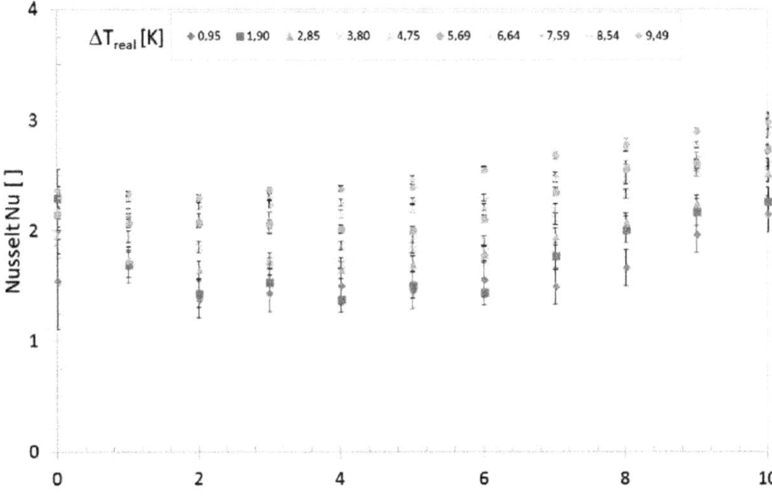

Figure 2.9.: The Nusselt numbers measured at different ΔT and voltages in the laboratory (Futterer, Dahley, *et al.* 2016).

3. Theoretical background

This chapter introduces some of the mathematical background. More information on the basic calculations and assumptions can be found in e.g. Pohl 1978, Yoshikawa *et al.* 2013, Meyer, Jongmanns, *et al.* 2016 and Meyer, Crumeyrolle, *et al.* 2018. Depending on the group which worked on this topic, different numbers are used to present the data. This makes it more difficult to compare our data to earlier publications. I will use either the physical units or the dimensionless numbers defined by Yoshikawa (Yoshikawa *et al.* 2013).

3.1. Thermo electro hydrodynamic force

The electro hydrodynamic force is described by (Landau *et al.* 1984)

$$\vec{F} = \rho_e \vec{E} - \frac{1}{2} E^2 \vec{\nabla} \varepsilon + \vec{\nabla} \left[\frac{\rho}{2} \left(\frac{\partial \varepsilon}{\partial \rho} \right) E^2 \right] \tag{3.1}$$

with E the electric field and ε the dielectric constant.

The first term describes the electrophoretic or Coulomb force F_C. This force is dominant when dc voltages are used, but can be neglected if an alternating field with a frequency higher than the inverse of the charge relaxation time is used. The charge relaxation time is defined as $\tau_e = \frac{\varepsilon}{\sigma}$. (Turnbull 1969) For silicone oils, which are used in this work, this time is in the range of several seconds, since the electric conductivity is very low $\sigma = 10^{-13}\frac{S}{m}$. Further also the viscous relaxation time $\tau_v = \frac{d^2}{\nu}$ plays a role in the choice of minimum frequency. The faster the change in polarization of the field happens compared to those two times, the better the result. The third term represents the electrostrictive force F_{ES}. Since the used silicone oil is incompressible and the walls of the containment are rigid, this plays no dynamic role. This leaves the second term which is the dielectrophoretic force F_{DEP}.

Since the density and dielectric constant are temperature dependent, the following Boussinesq approximations are made to linearize the temperature dependency of the density and permittivity

$$\rho = \rho_0 \left(1 - \alpha \left(T - T_0\right)\right) \tag{3.2}$$

$$\varepsilon = \varepsilon_1 \left(1 - e \left(T - T_0\right)\right) \tag{3.3}$$

with T_0 the reference temperature, $\rho_0 = \rho(T_0)$ and $\varepsilon_1 = \varepsilon(T_0)$.

This leads to the formula for the dielectrophoretic force (Yoshikawa *et al.* 2013)

$$\vec{F}_{DEP} = \vec{\nabla} \left(\frac{\varepsilon_0 \varepsilon_r e \vec{E}^2 (T - T_0)}{2} \right) - \rho \alpha (T - T_0) \vec{g_e} \tag{3.4}$$

with E the electric field.

This force can be seen as artificial electric gravity force. In contrast to the natural gravity g, the electric gravity g_e acts on the permittivity of the fluid and not the density. For a cylindrical geometry g_e is defined as

$$g_e = \frac{\varepsilon_r \varepsilon_0 e}{\alpha \rho} \left(\frac{V_{rms}}{\ln(\frac{R_1}{R_2})} \right)^2 \frac{1}{r^3} \mathcal{F} \tag{3.5}$$

with R_1 and R_2 are the inner and outer radii of the annulus and r is the position between R_1 and R_2 at which the strength should be calculated. To keep this value comparable within this work $r = \frac{1}{2}(R_1 + R_2)$ unless stated otherwise. The voltages are usually given as peak voltage V_p in the scope of this thesis. It can be related to the root-mean-square voltage or effective voltage by $V_{RMS} = \frac{V_p}{\sqrt{2}}$. $\mathcal{F}$ includes the temperature gradient and the dependence of the dielectric constant on the temperature. It is defined by

$$\mathcal{F} = \frac{\gamma_e^2 \left[1 - \gamma_e \left(\frac{\Theta(r)}{\Delta T} \right) + \frac{1}{\log \eta} \right]}{\left[\log(1 - \gamma_e) \right]^2 \left(1 - \frac{\gamma_e \Theta(r)}{\Delta T} \right)^3} \tag{3.6}$$

where $\gamma_e = e \Delta T$ is the temperature dependent dielectric constant and $\Theta = \frac{\Delta T \log(\frac{r}{R_2})}{\log \eta}$ describes the base temperature gradient in the gap.

3.2. Governing equations and base state

For this application the Navier-Stokes equation for incompressible Newtonian fluids with electric force field, which also includes the Boussinesq approximation, is used

$$\rho_0 \left(\frac{\partial \vec{u}}{\partial t} + \left(\vec{u} \cdot \vec{\nabla} \vec{u} \right) \right) = -\vec{\nabla} p + \mu \nabla^2 \vec{u} + \vec{F}_E - \rho_0 \vec{g} \alpha \left(T_{in} - T_{out} \right) \tag{3.7}$$

with ρ the density, $\vec{u}$ the velocity field, p the pressure, μ the dynamic viscosity, g the gravitational force, $\vec{F}_E$ the electrohydrodynamic force and α the thermal expansion coefficient. For the temperature difference $T_{in} > T_{out}$.

The continuity equation is given by

$$\vec{\nabla} \cdot \vec{u} = 0 \tag{3.8}$$

Since no heat sources are considered, the energy conservation can be described by

$$\frac{\partial T}{\partial t} + \vec{u} \cdot \vec{\nabla} = \kappa \nabla^2 T \tag{3.9}$$

with κ the thermal diffusivity.

The electric field which is applied is described by Gauss' law, which includes a linear approximation of the permittivity

$$\vec{\nabla} \cdot \left[(1 - \gamma_e T) \vec{\nabla} \Phi \right] = 0 \tag{3.10}$$

which includes the dimensionless temperature gradient

$$\text{where } \gamma_e = e\Delta T \tag{3.11}$$

with Φ the electric potential and e the thermal coefficient of permittivity, which is for the used fluids in the range of $10^{-3} \cdots 10^{-2}\frac{1}{K}$ (Yoshikawa *et al.* 2013).

3.3. Characterization by dimensionless numbers

The experimental parameters are given in dimensionless parameters to make a comparison between different fluids and experimental parameters. However, there are many possibilities and different groups chose different numbers. These are introduced here to make a comparison possible. To characterize the geometry the aspect ratio

$$\Gamma = \frac{H}{R_2 - R_1} \tag{3.12}$$

and radius ratio

$$\eta = \frac{R_1}{R_2} \tag{3.13}$$

are used. Depending on these ratios the number of vortices and the critical point at which they occur can change.

The diffusion properties is described by the Prandtl number

$$\mathcal{P}r = \frac{\nu}{\kappa} = \frac{\text{viscous diffusion rate}}{\text{thermal diffusion rate}} \tag{3.14}$$

with ν the kinematic viscosity and κ the thermal diffusivity. For low Prandtl numbers ($\mathcal{P}r < 1$) the heat transfer via conduction is more dominant than via convection, while for high Prandtl numbers it is the other way around.

To determine the Archimedean buoyancy, the Grashof number is used

$$\mathcal{G}r = \frac{g\alpha\Delta T\left(R_2 - R_1\right)^3}{\nu^2} = \frac{\text{buoyancy force}}{\text{viscous force}} \tag{3.15}$$

When the Grashof number reaches a critical point, Gr_c, instabilites start to occur. Since this number include g, it can only be applied when either g or g_e acts on the fluid. A superposition of both forces cannot be resolved by this number.

From these numbers the Rayleigh number is derived

$$\mathcal{R}a = \mathcal{G}r\,\mathcal{P}r = \frac{g\alpha\Delta T\left(R_2 - R_1\right)^3}{\nu\kappa} \tag{3.16}$$

Below a certain critical number, $\mathcal{R}a_c$, the heat is transferred primarily via conduction, while above the critical point it is primarily via convection.

Analogous to $\mathcal{R}a$ an electric Rayleigh number $\mathcal{R}a_e$ is defined by replacing g with g_e

$$\mathcal{R}a_e = \frac{g_e\alpha\Delta T\left(R_2 - R_1\right)^3}{\nu\kappa} \tag{3.17}$$

Since the strength of the electric field is depending on the position inside the gap, $\mathcal{R}a_e$ is defined to be the middle of the gap.

To quantify the convective heat transfer the Nusselt number is used. Depending on the setup of the experiment a different formula is used. In both cases the Nusselt number is defined as $\mathcal{N}u = \frac{\text{total heat transfer}}{\text{conductive heat transfer}}$ (Bejan 2004 and Martynenko *et al.* 2005). If $\mathcal{N}u = 1$, then all heat is transferred by conduction. The higher the Nusselt number the higher is the convective heat transfer compared to the conductive transfer.

$$\mathcal{N}u = \frac{P_{heating}}{\frac{\lambda A_{in} \Delta T}{R_1 \ln(R_2/R_1)}} \tag{3.18}$$

with $P_{heating}$ the heating power put into the system on the inner cylinder, ΔT is the measured temperature difference between inner and outer cylinder, λ the thermal conductivity and A_{in} the outer lateral area of the inner cylinder. This formula is used for early experiments at our department performed by e.g. Dahley (Dahley 2014 and Futterer (Futterer, Dahley, *et al.* 2016). The maximum heating power $P_{heating}$ was measured and could be regulated in the range of $0 \ldots 100\%$.

After the design of the cells was changed to a heating liquid loop the calculation of the total heat input in the system was changed to:

$$\mathcal{N}u = \dot{V} c_p \left[\rho(T_{h,i})T_{h,i} - \rho(T_{h,o})T_{h,o}\right] \left(\frac{2\pi H \Delta T}{\ln \frac{R_1}{R_3}/\lambda_{Al} + \ln \frac{R_2}{R_1}/\lambda_{oil} + \ln \frac{R_4}{R_2}/\lambda_{Al,Glass}} \right)^{-1} \tag{3.19}$$

where $\dot{V}$ is the volume flow rate of the heating loop in m^3/s, T the temperature in K. The subscripts h and c refer to the heating and cooling loop and i and o describe whether it is measured at the inlet or outlet. $\rho(T)$ the temperature dependent density in kg/m^3, c_p the specific heat capacity of the fluid in $J/(kg \cdot K)$, λ the thermal conductivity in $W/(K \cdot m)$, H the height of the cell in m, and $\Delta T = T_{h,i} - T_{c,i}$. The thermal conductivity changes depending on the cell type. The inner cylinder is always made of aluminium, but the outer cylinder can be made of glass or aluminium. The radius R_1 describes the outer wall of the inner cylinder and R_3 the inner wall of the inner cylinder. The radii R_2 and R_4 describe the inner and outer wall of the outer cylinder respectively. Figure 3.1 gives an overview of these parameters.

Instead of the voltage the dimensionless potential (Yoshikawa *et al.* 2013) is used

$$V_E = \frac{V_{rms}}{\sqrt{\frac{\rho \nu \kappa}{\varepsilon_0 \varepsilon_r}}} \tag{3.20}$$

The earlier publications usually non-dimensionalize the temperature gradient as $\mathcal{R}a$ and the voltage as $\mathcal{R}a_e$. The problem is that $\mathcal{R}a_e$ is a function of V_p and ΔT. For this reason Yoshikawa uses γ_e to display the dimensionless temperature gradient and V_E as dimensionless potential. This way there is a dedicated number for each variable, ΔT and V_p, and makes a comparison between the LSA and experiments easier.

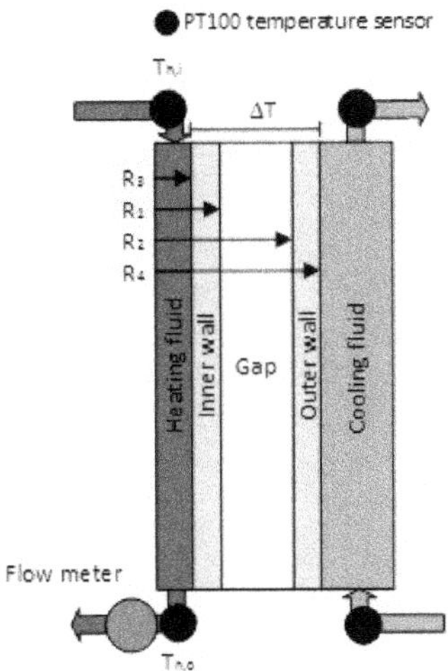

Figure 3.1.: Schematic presentation of some parameters needed to calculate the Nusselt number

4. Experiments in microgravity

The parabolic flight is one of several ways to create a microgravity environment. Table 4.1 gives a rough overview over the characteristics of several possibilities. While the parabolic flight doesn't offer a very good μg quality compared to the other methods, it is quite easy to access.

Table 4.1.: Four possible platforms for low gravity research (ESA 2014)

	μg quality in g	μg duration	required preparation time
Drop Tower	10^{-6}	$9.3s$	3 to 6 months
Parabolic Flight	10^{-2}	$22s$	3 to 9 months
Sounding Rocket	10^{-4}	6 to 13 min	12 to 24 months
Space Station	10^{-6}	continuous	9 to 60 months

Sounding rocket flights are very limited in availability and may not be available every year. During this flight the experiment has to run autonomously and all parameters have to be prepared very carefully. The sounding rocket offers a good μg quality, but it requires a very good and precise preparation, which cannot be reached by only laboratory experiments. So other possibilities, such as drop tower experiments or parabolic flights, have to be used in preparation. The size of the experiment is very limited, since it needs to fit inside the loading area of the rocket together with other experiments.

Access to the space station ISS is even more restricted and the preparation time and cost for such a project are very high. If one wants to put an experiment on the ISS a very long and careful preparation time has to be considered. Similar to the sounding rocket the space and other resources, such as available power, are limited. While the μg quality is very good and the experiment can run for a long time, it is not a realistic option due to the required preliminary work.

The drop tower offers a very good μg quality and it is not difficult to apply to an experimental run. At the end of the drop the experiment capsule is stopped rather abrupt, which means that the experiment needs to be designed sturdy enough to not be damaged by this. The biggest drawback is the rather short μg time of only $9.3s$, when the catapult system is utilized. In theory, this is enough time to see changes in the flow of our experiment, because $9,3s > \tau_v$. But, depending on the parameters, the growth rates of the perturbations can be slow and it would require more time.

The parabolic flight has the worst μg quality of these options. The time is also quite limited, but longer than in the drop tower. To get on a parabolic flight campaign one needs to apply at the German (DLR), French (CNES) or European (ESA) space agency.

The accessibility is also limited by the number of flights the space agencies organize and the experiment has to be chosen by a committee. One of the big advantages of these flights is that the resources are not as limited as in the other options. There is more space, making it easier to construct the experiment. It is also possible for the experimenter to be in the aircraft during the flight. If something does not work as planned it is still possible to manually correct the experimental parameters during the flight. Since the experimenters are confined in a aircraft during the experimental runs, special security requirements are in place, which makes the construction of such an experiment more challenging and restrictive than a laboratory experiment.

4.1. The parabolic flight campaign

For European users the parabolic flight campaign is organized by the french company Novespace in Bordeaux, France. They maintain a modified Airbus A310 aircraft, which is used during the campaign. They also perform all necessary safety checks for the experiments and give advice to the participants.

The campaign not only includes the flight, but also the guidance by Novespace while preparing the experiment. This starts about 6 months before the actual flight. During this time the experimenters and Novespace employees are in contact to discuss the state of the experiment. Novespace checks if the experiment complies to the safety guidelines and the experimenters have to adapt the experiment for the flight.

When this process is done, the experiment is shipped to Novespace. The experimenters and their experiments arrive at Novespace the week before the flight. During this week the experiment is again tested and examined by their technicians. After approval the experiment is loaded into the aircraft. After loading, it is checked a last time by members of the DGA essais en vol. The DGA essais en vol is an organization responsible for the military and civil use of aircrafts. They also provide the pilots for the flight. It is possible to work with the experiment and make last tests and calibrations until the end of the week. The Monday of the flight week is usually for the safety briefing for the people participating in the flight. These are the people operating the experiments, but also test subjects of physiological or biological experiments. Tuesday to Thursday or Friday, depending on the number of flight days, are the flight days with each a flight of about 3 hours and a debriefing after that. On the afternoon it is possible to make modifications to the experiment for the next day. After the last flight the experiments are immediately unloaded from the aircraft and need to be packed and shipped back to the original facility.

4.2. Parabolic flight

Each of the flight days starts with an inspection of the experiment inside the aircraft. It is checked if e.g. oil is leaking and the optical access to the cells is still clean. The people assigned to fly this days will get a scopolamine injection from a doctor. This drug inhibits the effect of motion sickness, which can be induced by the changing gravity phases during the flight, but also causes some side effects like drowsiness and a drying of the mucous membranes. The doctor is also present during the flight to check on the participants.

After take-off the aircraft travels to the target location where the parabolas are performed. This is usually above the Atlantic Ocean near the coast of France. But if the weather conditions are not acceptable the flight will be performed above the Mediterranean Sea. Each flight consists usually of 31 parabolas. The first one, parabola 0, is used as warm-up parabola for the test subjects and to look out for loose items which are flying in the cabin. The parabolas are divided into 6 sets. The first one has 6 parabolas and the others 5. Between these sets is a pause of 5 to 10 minutes. In these pauses it is possible to make adaptions to the experiment.

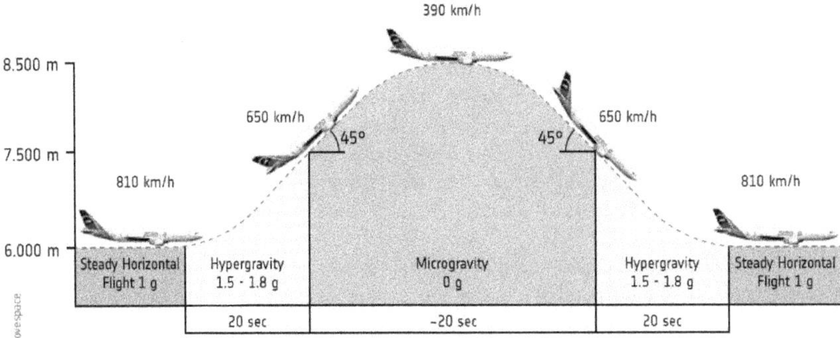

Figure 4.1.: Sequence of one parabola during a parabolic flight (ESA 2014). The aircraft decelerates and pulls up, which increases the gravity to about $1.8g$. At a certain angle the μg phase starts. After that the aircraft accelerates and pulls into normal flight position, which creates a $1.8g$ phase again.

The maneuver flown during a parabola can be seen in Figure 4.1. The aircraft begins in a steady flight state at $6000m$ altitude and a velocity of $810km/h$. This phase is comparable to a normal commercial flight, but at lower altitude. The aircraft is first accelerated and pulled up. During this pull-up phase the aircraft is decelerated and there is a hypergravity of about $1.8g$. At an altitude of about $7500m$ the aircraft has an angle of elevation of $\geq 47°$. This is when the μg phase begins. The engine power is throttled even further to reduce the air-drag and the aircraft follows a ballistic trajectory. After about $20 \ldots 23s$ the power of the engines is increased again and the aircraft is pulled back into a steady phase. During this pull-out phase there is again a hypergravity phase of about $1.8g$. This whole maneuver from steady flight to μg to steady flight takes about 70 seconds.

When all parabolas are completed the aircraft returns to Bordeaux-Merignac airport. After landing there is usually a break so the participants can get some rest and have lunch. About 1 hour after landing there is a debriefing. During this every participant has to state how their experiment went and whether problems occurred or not. Then there is some time to prepare the experiment for the next day and make a backup of the collected data. The experiment has to be in flight condition for the next day, before leaving the Novespace facility.

4.3. Restrictions to the experimental design

The design of the experiment has to follow certain guidelines given by Novespace. These are rather strict compared to the guidelines for laboratory experiments. To give an overview to Novespace what the experiment does, which risks could occur and how they are handled, the Experimental Safety Data Package (ESDP) is written. This document describes the purpose of the experiment, mechanical structure, electrical systems and all procedures relates to the handling of the experiment.

There are general rules for e.g. attaching the experiment to the aircraft and usage of electrical power. Additionally the experiment is rated more detailed depending on the used systems such as heating, cooling, fluid loops, gasses and so on. The potential risks for this experiment are given in table 4.2. Each of these has to be dealt with to prevent the hazards from happening or to limit the consequences and prevent any further damage.

Since the aircraft is a small, closed space, fire safety is very important. The fluids in the experiments should not have a low flash point or high flammability. For AK5, this is high enough, but for other fluids, such as AK0.65 it was more difficult to obtain approval to use it. The heating and cooling loops are always filled with AK5 and only the fluid in the gap changes. The amount is very small compared to the volume of the other fluid loops. In case anything happens, the working fluid mixes with the AK5, which lowers the flash point. This is a high risk factor, since the fluid is actively heated and a flashover by the high voltage could ignite the fluids. In case any fluid loop leaks, the Experiment Box acts as double containment and no fluid can get in the aircraft. The risk of a developing fire is also limited by the air supply in the box and the fire is stifled by the lack of oxygen.

The voltage and related electromagnetic radiation caused by the voltage is also rated high, but can be handled rather easily. All electrically conductive parts of the experiment are connected to protective earth (PE) of the electric system. This causes all currents induced by the electromagnetic interferences to be removed to the PE line. This should ideally happen over the high-voltage amplifier, which also causes these interferences. However, the current going over the PE line to the power outlet also increases slightly, which may trigger the ground fault interrupter fuse. The Experiment Box is made from aluminium, which is also connected to PE. This absorbs nearly all electromagnetic fields, which are induced by the systems inside the box. While the voltages are very high, up to $V_p = 10kV$, the currents are rather low. They are limited to a maximum of $I = 10mA$ by the high-voltage amplifier. This limits the potential damage in case of a short circuit.

All hot surfaces on and inside the Experiment Box needs to be marked and protected against contact. This is also required for moving objects, such as cooling fans. This is simply done by placing a grid on the Experiment Rack, which protects these places and components from possible contact. Most of these areas are located on the backside of the rack, which is placed against the wall of the aircraft and cannot be reached.

In case of a pressure loss in the cabin it needs to be ensured that the closed Experiment Box does not implode. This is done by placing a rubber disc in the Box, which ruptures in the event of a pressure loss and prevents any damage to the Experiment Box.

Table 4.2.: Risk assessment as given in the ESDP. The risk is classified as minor (operational limitations for the experiment), major (light injuries), critical (serious injuries, severe equipment damage) or catastrophic (aircraft equipment destroyed, fatal injury). Every item on this list needs 1 (major, critical) or 2 (catastrophic) technical barriers. These have to prevent any damage potentially caused by the risk.

Hazard Group	Hazardous conditions/Risks	Hazard classification
Radiation (Ionizing, electromagnetic, laser)	• Exposure to RF emission • Laser radiation	Critical Critical
Fire	• Overheating of heating system leading to fire • Electrical fire	Catastrophic Catastrophic
Electrical Shock/Static Discharge	• Electrical circuit and/or HV generator	Catastrophic
Structural failure	• Rupture of the rack during emergency landing	Catastrophic
Collision / Impact	• Injury with sharp edges	Major
Injury and/or Illness	• Contact w/ hot surfaces (inside box) • Contact w/ hot surfaces (outside Peltier) • Contact with fans • Eye injuries by Laser light (class 3b)	Critical Critical Major Critical
Explosion-Implosion	• Loss of cabin pressure leading to overpressure of Zarges boxes and liquid container	Critical
Loss of Habitable Environment	• Cabin contamination with spread of liquids or frangible material like PMMA, glass.	Critical

5. Experimental setup

For the purpose of this experiment we designed a dedicated setup. A first version of the experiment was developed at our department in 2010 (Dahley 2014). More than 4 students wrote their bachelor or master thesis on this topic. For my first PFC in 2015 I used their setup. Some changes were made based on preceding laboratory experiments. Later it was completely re-designed.

Since this experiment is used in parabolic flights, the structure needs to comply to certain rules given by Novespace. One of their suggested setups for a complex experiment such as ours is to use two racks (Figure 5.2). One is used to control the experiment and the other includes the experiment itself. The advantage is that it is easier to handle than one rack, since the weight is distributed on two racks and it is easier to place two small racks into the aircraft than one big rack. Also, the Experiment Rack can be placed on the side directly at the wall of the experiment area inside the aircraft, while the Control Rack is positioned more accessibly by the experimenters near the alleyway in the middle of the area.

During the time I worked on this thesis we designed two different Experiment Box setups within this project and one more with a different geometry for the university in Le Havre. The first Experiment Box is used to measure the temperature at several points of the outer cylinder of the cell and uses Shadowgraph or Synthetic Schlieren method for the visualization. The second box is used for LLS/PIV and uses fewer temperature converters. The third one has the possibility to do either Shadowgraph or LLS/PIV and uses cells with a rectangular cavity. The results of this third Experiment Box are not included in this thesis.

Both, the Experiment Box and the control PC, were mounted on one rack (Figure 5.1). While the used cell geometry was the same, some aspects of the design of the experiment cells were different.

The used experiment cells had a height of $H = 100mm$, gap width of $d = 5.1mm$ and radius ratio of $\eta = 0.5$. To apply the temperature gradient, the cells were cooled by a fluid loop on the outside and heated by an electric heating cartridge on the inside. Temperatures were measured at six positions on the outer cylinder and the Shadowgraph technique was used for visualization. A more detailed description can be found in the dissertation of Norman Dahley (ibid.).

This setup was heavily modified for several reasons. One is that the cell design included glue and adhesive compounds. The glue prevented the exchange of single components of the cell and made it impossible to clean it. The heating was done with a heating cartridge inside the cell, which allows a fast and precise control of the temperature, but also caused negative interactions with the applied voltages. The mechanical structure was also changed. Instead of one rack for the Experiment Box and the Control PC,

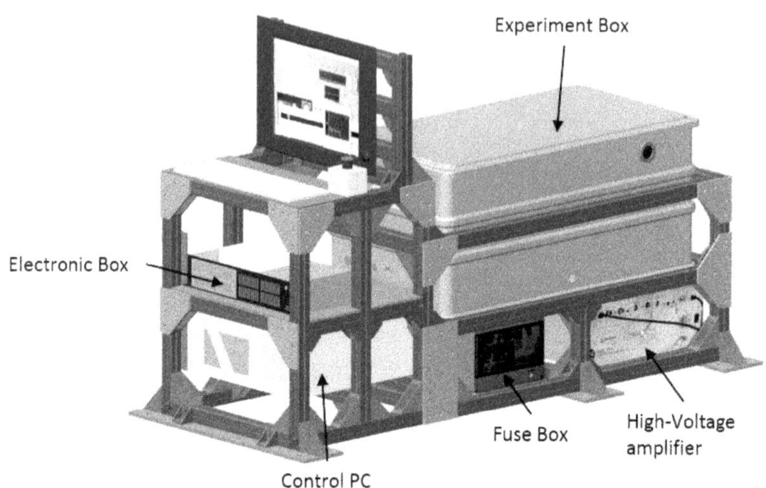

Figure 5.1.: The previous experimental setup (Dahley 2014). The experiment cells are inside the Experiment Box. The high-voltage amplifier, Fuse Box and Electronic Box are used to supply the experiment with power and control the components, such as heating and cooling. The PC is used to control the experiment and store the data via a LabVIEW interface.

two racks are used. This setup is easier and more flexible to handle inside the aircraft. Since the Experiment Box can be changed easily within one hour, we can use different experimental setups during one PFC.

5.1. Re-designed setup

Specialized experiment cells were designed by our group specifically for this application. The cells need to have an active cooling and heating to create a temperature gradient. No silicon-based compound material may be used, since the silicone oil inside the fluid loops would dissolve these components and because of the applied voltages most of the used materials need to be electrically isolating. These restrictions pose a challenge to the design of the cells.

Figure 5.2.: The control rack (left) and experiment rack (right) of the PFC setup constructed in 2016.

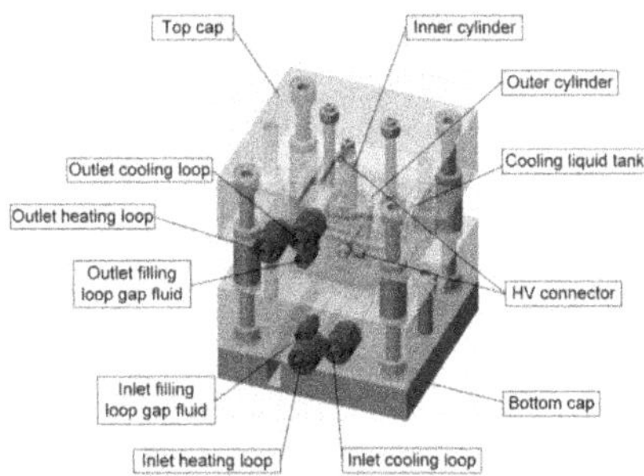

Figure 5.3.: CAD drawing of the 30mm cell.

Two different types of experiment cells have been designed to be used with different measurement techniques. The inner cylinder is made from aluminium, because it is a good thermal ($\lambda = 210\frac{W}{m \cdot K}$) and electric conductor. The material of the outer cylinder depends on the used measurement technique (Table 5.1). For the first type of experiment cells, Type T, also aluminium is used. Several holes are drilled into the cylinder and thermoucouples type J are placed inside them. They are glued in place using a thermal-conductive compound. This way we can measure the temperature on several locations at the outer cylinder. There are up to 24 sensors on the $100mm$ cells and up to 22 on the $30mm$ cells. They are divided in two columns with 5 ($30mm$ cells) or 6 ($100mm$ cells) sensors each. In addition there are two rows of 8 sensors each near the top and bottom of the cell (Figure 5.4). The other type of experiments cells, type P, have outer cylinders made of borosilicate glass ($\lambda = 1.2\frac{W}{m \cdot K}$). No thermocouples are mounted inside these cylinders. Since the cylinders need to be electrically conductive they have been coated with a transparent conductive oxide (TCO). The coating absorbs nearly no light in the visible spectrum, so that it is possible to apply a LLS technique to those cells.

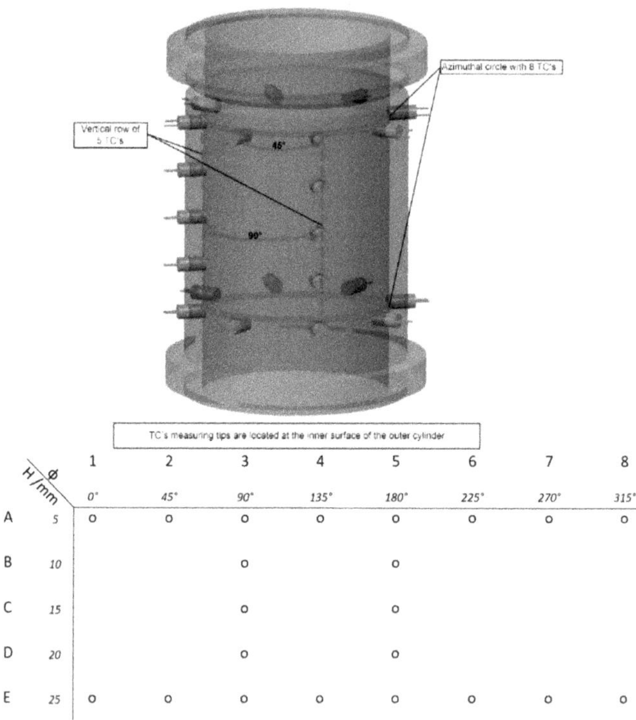

	H/mm \ ø	1 0°	2 45°	3 90°	4 135°	5 180°	6 225°	7 270°	8 315°
A	5	o	o	o	o	o	o	o	o
B	10			o		o			
C	15			o		o			
D	20			o		o			
E	25	o	o	o	o	o	o	o	o

Figure 5.4.: Position of thermocouples on the cell with 30mm height. There are two columns with 5 thermocouples in each of them. The columns are located 90° from each other. Near the top and bottom are two rows of thermocouples with 8 sensors in each.

In this modified version of the experiment cells, the heating at the inner cylinder is realized with a heating fluid loop. The loop is filled with silicone oil AK5. There

Table 5.1.: Properties of the experimental cells. The used aluminium is an AlMgSi0.5 alloy.

	Type T cells	Type P cells
Inner cylinder	Aluminium	Aluminium
Outer cylinder	Aluminium	Borosilicate glass
Temperature sensors	$22\dots24$	0
Measurement techniques	Shadowgraph Synthetic Schlieren	PIV

is a heating box which includes two heating cartridges to heat the fluid. The fluid is pumped into the inner cylinder, then into an expansion tank to compensate for the expansion of the fluid when it is heated and back to the heating box. Inside each of the heating cartridges is one thermocouple to monitor the temperature. Additionally two self-resetting thermal cutoffs are connected to the outside of the heating box. While the thermal cutoffs are a hard limit for the maximum temperature, the temperature from the thermocouples is used to implement a shutdown in the control software. Both systems are required by the safety standards of a parabolic flight. The maximum temperature inside the heating loop is limited to $65°C$. At each, the inlet and outlet of the heating loop, is one thermocouple to determine the temperature inside the loop. These sensors are used in all cell types. The flow-rate inside the heating loop is measured using a flowmeter, which has been calibrated for the use with AK5 and also includes the temperature dependent density of AK5.

Similar to the heating fluid loop, there is also a cooling fluid loop. This loop is also filled with silicone oil AK5. But here heat exchangers with Peltier elements are used to control the temperature. The Peltier elements are mounted on the wall of the Experiment Box. On the outside of the box are several heat sinks and fans to remove the heat. For safety reasons there are also thermal cutoffs and a software cutoff, if the temperature exceeds $65°C$. I build a dedicated controller for the Peltier elements, so that the fluid can be cooled or heated and set to a constant temperature (Appendix C). The temperatures at the inlet and outlet of the cooling loop at the experiment cell are also measured by thermocouples.

To assist in filling and changing of the working fluid inside the gap there is a third fluid loop including pump, reservoir and expansion bellow. During operation of the experiment this fluid loop is can also be used to mix the particles, which are needed to apply PIV.

The outer boundaries of the experiment cell, i.e. the walls outside the cooling fluid loop and top and bottom caps, are made from polymethylmethacrylate (PMMA) with a heat transfer coefficient $\lambda = 0.19\frac{W}{m \cdot K}$, $8mm$ thickness of the walls and $45mm$ thickness of the caps. Due to the low heat transfer coefficient and the thickness of the walls we assume an adiabatic system, which means that all heat energy, which enters the cell, is transferred via the gap and only a negligible amount is lost to the environment.

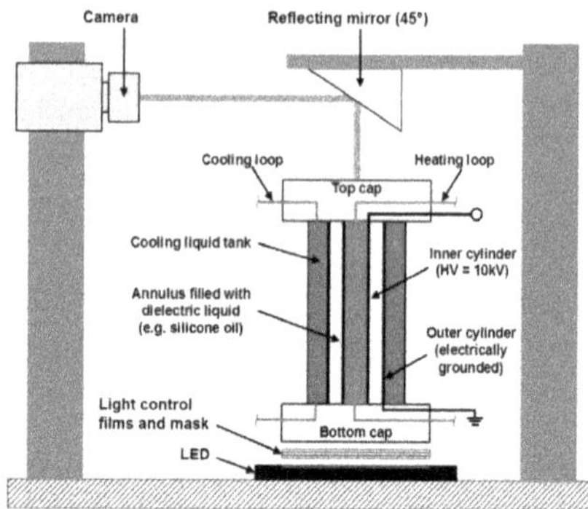

Figure 5.5.: The optical setup for the Shadowgraph and Synthetic Schlieren methods (Dahley 2014). A red LED panel is placed below the cell. The light is rectified by light control films, which produce quasi-telecentric light. For the Synthetic Schlieren method a mask with a pattern is placed on top if these films. The light is redirect by a mirror to the camera.

The Experiment Box type T (Figure 5.7) is used for measurements of the temperatures on the outer cylinder (Figure 5.4) and visualization of the flow inside the gap by Shadowgraph or Synthteic Schlieren methods. Up to 64 thermocouples have to be measured and an appropriate number of measurement converters is required to do this. All converters have isolations between the inputs and output. This protects the measurement hardware and the PC in case of a flashover of the high-voltage. Long cables on the thermocouples can also increase the error slightly, which can be prevented by measuring the signals inside the box close to the thermocouples. The gap of each cell is illuminated by a red LED-array, which is located below the cell. To create quasi-telecentric light light control films (LCF) are used. In addition a mask is added, when the Synthetic Schlieren method is applied (Figure 5.5). The images are captured by a camera (Table A.1) at $10Hz$ and a resolution of $1600x1200$ pixels.

The Experiment Box type P (Figure 5.8) is used for measurements with the LLS and PIV methods. A line laser is aligned to the cell in a way that the light sheet spanned by the laser illuminates a region along the axis of the gap (Figure 5.6). It is a continuous red laser with a wavelength of $\lambda = 620nm$ and power of $100mW$. The camera (Table A.1) is mounted perpendicular to the laser plane. The images are captured as $1920x1200$ grayscale images at $10Hz$. To apply the PIV evaluation method tracer particles are required inside the fluid in the gap. For AK5 we used Potters hollow glass spheres with $\rho \approx 1g/cm^3$ and for AK0.65 Thermo-Spheres W with $\rho \approx 0.8g/cm^3$. A first study for suitable tracers was done by Norman Dahley (Dahley 2014) and he used polyamide particles. These had the drawback that they would slowly agglomerate when the high-voltage field was applied. This study was continued by our group. Because of the high

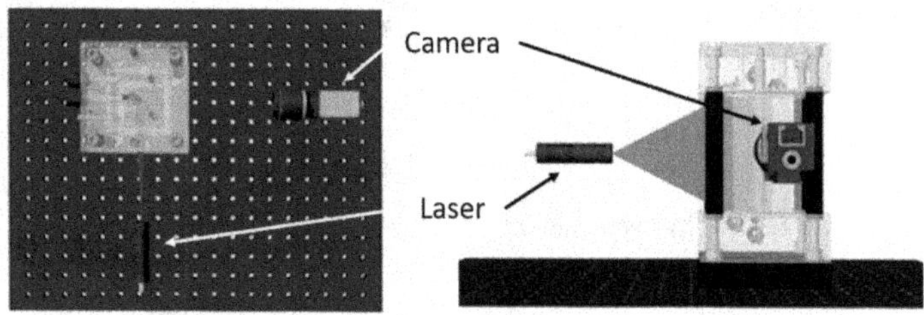

Figure 5.6.: The optical setup to apply LLS/PIV (Meier *et al.* 2018). The laser is mounted parallel to the axis of the cell. The camera is placed perpendicular to the laser plane.

voltage, the choice of particles is limited. They need to be around the same density and dielectric constant as the fluid and need to be made of electrically non-conducting material.

Both Experiment Boxes are controlled from the PC on the Control Rack. The PC has two National Instruments (NI) Data Acquisition (DAQ) cards. Up to 70 analog values are measured and several analog and digital outputs need to be generated. This is done by a NI LabVIEW program (Appendix B), which I developed for this project. There is also an accelerometer to measure the acceleration forces acting on the experiment. The data from it can be accessed from LabVIEW, so that the whole program flow and experimental settings can be synchronized to the individual gravity phases of the parabolas.

More detailed technical information about the used hardware can be found in appendix A.

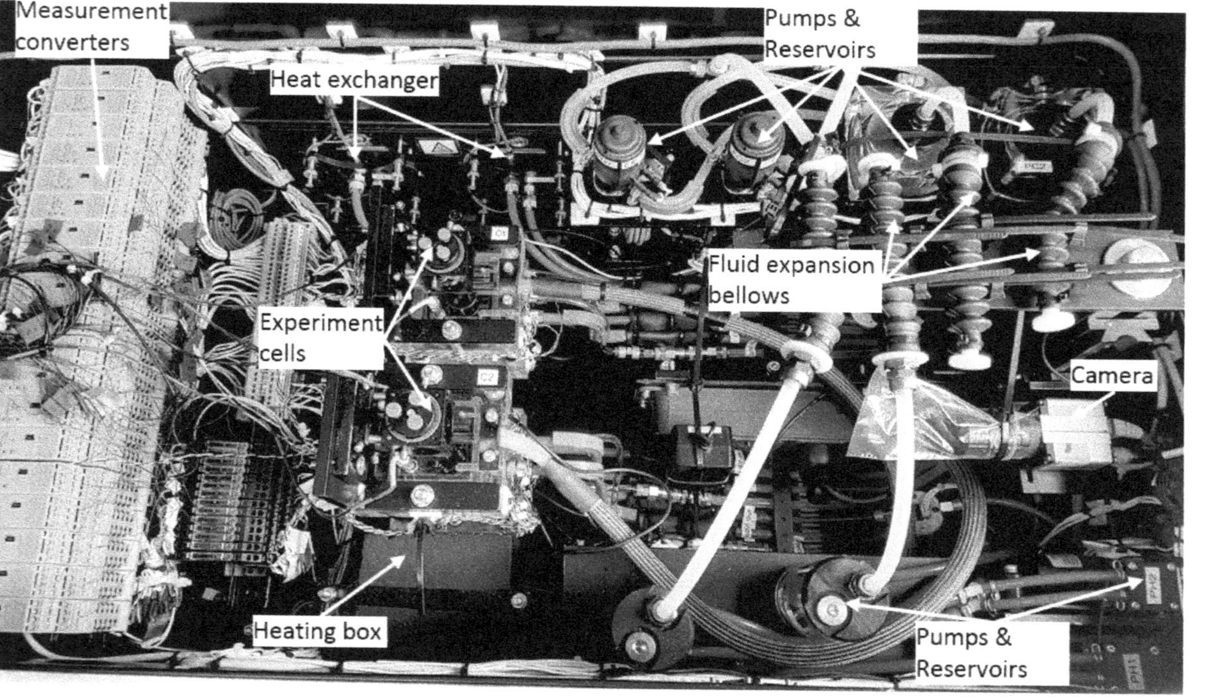

Figure 5.7.: A photo of the actual setup inside the type T Experiment Box. Each experiment cell is connected to three fluid systems, one for heating, cooling and filling/mixing. One heating and cooling system is present per cell. Due to the amount of thermocouples (up to 64 for both cells) a lot of measurement converters are required. Below the cells are LED arrays. The light is redirected by mirrors to the cameras.

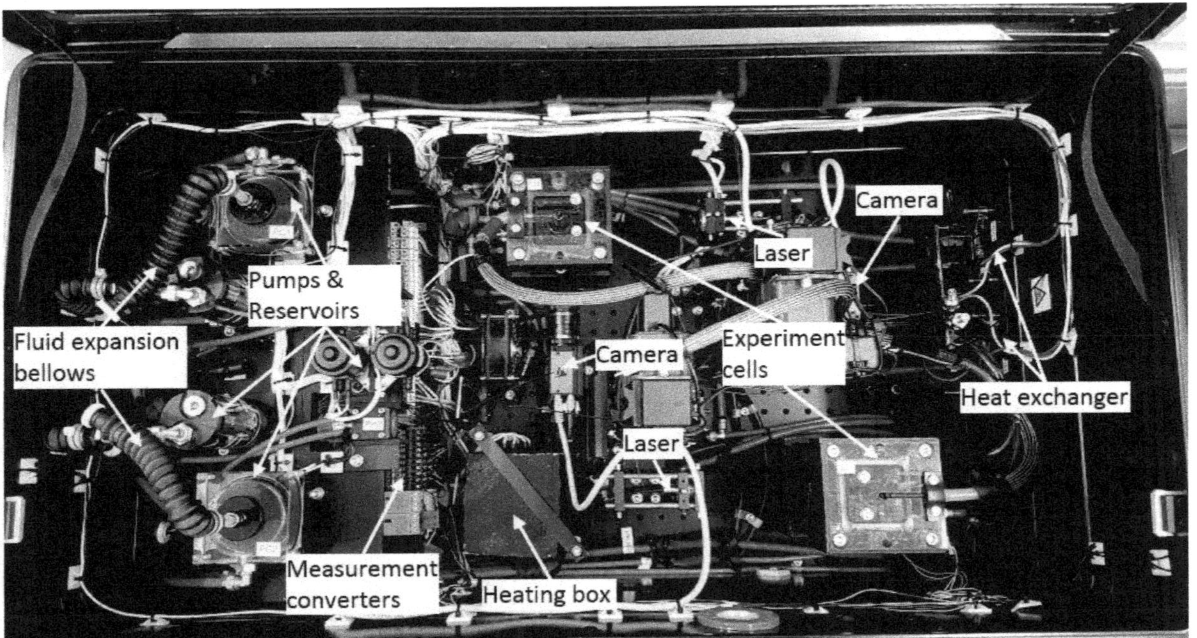

Figure 5.8.: A photo of the actual setup inside the type P Experiment Box. Each experiment cell is connected to three fluid systems, one for heating, cooling and filling/mixing. One heating and cooling system is present per cell. The number of measurement converters is lower than in the T Box, since no temperatures are measured on the outer cylinder. Instead of a LED array, a laser is required to perform LLS.

6. Measurement techniques

For this experiment we tested and qualified several measurement techniques to use during the parabolic flights. There is only a limited time during the parabolas, so the techniques have to be able to produce results within seconds after a change occurs. Further, the electric field limits the possible measurement methods. Because of flash overs, it is not possible to include probes in the gap of the experiment. Thus non-invasive methods are preferred.

The temperature measurements, which are required to determine the heat transfer, are done at the inlets and outlets of the fluid loops. They are outside of the electric field and show no interferences when the voltage is active. Additional measurements have been done on the outside of the outer cylinder to determine the temperature field. However, heat transfer is a rather slow process and it is not possible to reach a stable heat transfer state within the time of one μg phase. However, this data can be used as trend in which direction the temperature and heat transfer, given by $\mathcal{N}u$, develops.

The shadowgraph method (Schöpf *et al.* 1996) was already used by Futterer and Dahley in the preceding project and it proved to be usable for this experiment. However, under certain circumstances I was not able to get good results with this method. With a slight modification of the setup of the Shadowgraph method it was possible to use the Synthetic Schlieren method (Dalziel *et al.* 2000, Raffel 2015). I did several test under different conditions to determine the optimal usage of these two methods for the used experimental and geometrical parameters.

Another applied visualization is a Laser-Light Sheet (LLS) approach with Particle Image Velocimetry (PIV). While this has been used by Futterer and Dahley as well, there have been several problems. The particles need to have the same density and dielectric constant as the fluid. Further, they need to be made of an electrically non-conducting material and also should not be influenced by the dielectric force in any way. These requirements makes it difficult to find fitting particles. We continued based on their experiences and optimized the application of this technique further.

6.1. Temperature measurements

To test and qualify the used setup for the parabolic flight, the performance of the used thermocouples and measurement converters had to be quantified.

The most important characteristics are the precision and accuracy of the system. Each component used to measure the temperature has an influence on the error of the measurements. There can be two types of error. The systematic errors are static errors caused by the setup of the system and the used components. This type of error reduces the accuracy of the system, which determines how close the measured temperature is to the real temperature. It can be removed by performing calibrations using a very accurate reference.

The precision describes the spread of the measured temperature between several experimental runs with the same parameters. The results are expected to be exactly the same. However, this will almost never be the case. It depends on external influences, such as the environment temperature, which can influence the measurement hardware, or the presence of electromagnetic interferences. This type of error cannot be predicted and is hence called random error. It can be diminished by applying statistical methods on the data set, such as calculating a moving average and other filters.

To calibrate the sensors, I connected the heating and cooling loop of the experiment cells in series. The reference temperature was measured in the middle of the fluid loop with a Hanna Instruments Checktemp 1 thermometer. This is done to offset the heat losses or gains in the tubes, which connect the experiment cell to the thermostat. All metal parts of the fluid system were thermally insulated. This way I can assume that the temperature inside the experiment cell is the same as at the reference measurement. If I would have taken the reference measurement directly at the thermostat, then there would be an additional, temperature-dependent error of up to $1.4K$ caused by the uninsulated connecting tubes. The temperature of the fluid was controlled by an Anton Paar Viscotherm VT2 thermostat. The following temperatures were used as reference points: $20°C$, $25°C$, $30°C$, $35°C$, $40°C$ and $45°C$. The temperatures were set at the thermostat and maintained for 1 hour to allow the system to become thermally stable. For each reference and temperature I calculated the mean value over the last 10 minutes of the data. The uncalibrated data is plotted against the reference temperature (Figure 6.2) to show the initial situation. All temperatures are above the reference measurement with an error of up to $2.5K$.

The calibration is done in two steps. In a first step, a linear polynomial is calculated from the discrete values of the measurements (Fig. 6.1). The measured value from the thermocouples is given on the x axis and the reference temperature on the y axis. This results in a linearization of the measured temperature in the form of $T(T_{measured}) = a \cdot T_{measured} + b$. In a second step the error between the reference temperature and thermocouple temperature is calculated. This is done by calculating the difference of each of the discrete points $Error = T_{measured} - T_{reference}$. Again, a linear polynomial is calculated to give the error as function of the measured temperature $Error(T_{measured}) = a_{error} \cdot T_{measured} + b_{error}$. Now, both a and b coefficients can be added $a_{final} = a + a_{error}$ and $b_{final} = b + b_{error}$. The result is the function $T_{calibrated}(T_{measured}) = a_{final} \cdot T_{measured} + b_{final}$ which calculates the actual temperature, given by the reference temperature, as function of the measured temperature. This approach improves the accuracy from $2.5K$ down to $< 0.2K$ (Figure 6.3).

The reaction time of the measurement system is also important, since the experimental runs during a PFC are very short. Depending on the size of the thermocouple this time can be several seconds. For this reason we used thermocouples with a diameter of only $0.08mm$ by Omega Engineering. They have a low heat capacity and react fast to changes of the temperature. The compound (OmegaBond by Omega Engineering) used to attach

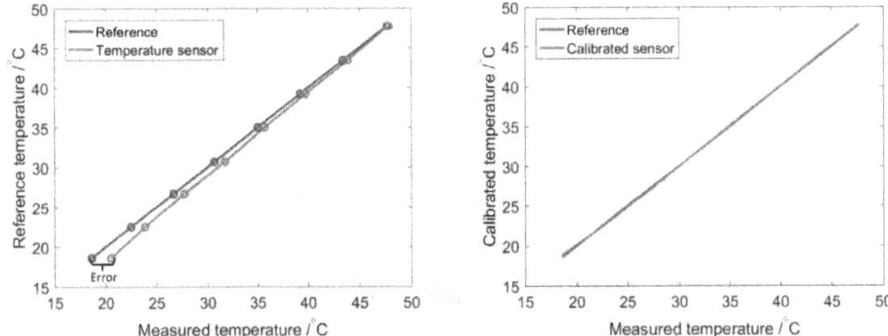

Figure 6.1.: Left: The discrete values of the measurement (circles) and a linear polynomial are shown. The difference of the discrete values of the reference and the sensor is used to calculate a linear, temperature-dependent error. Right: The same data after the calibration process. The temperature from the sensor has nearly the same linearity as the reference.

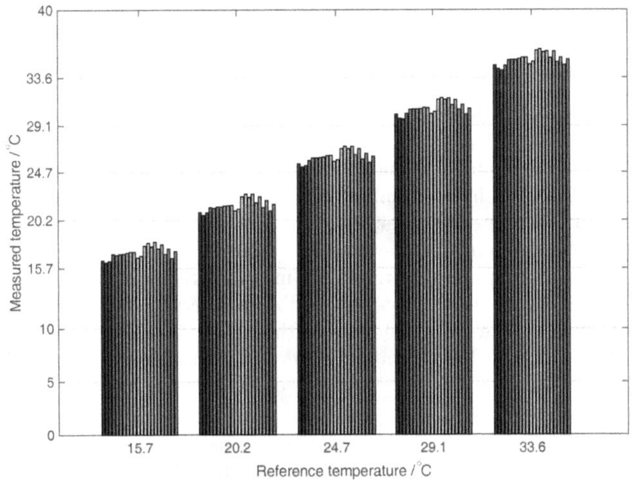

Figure 6.2.: Measured temperatures of the thermocouples without calibration compared to the reference measurement. The absolute error is up to $2.5K$.

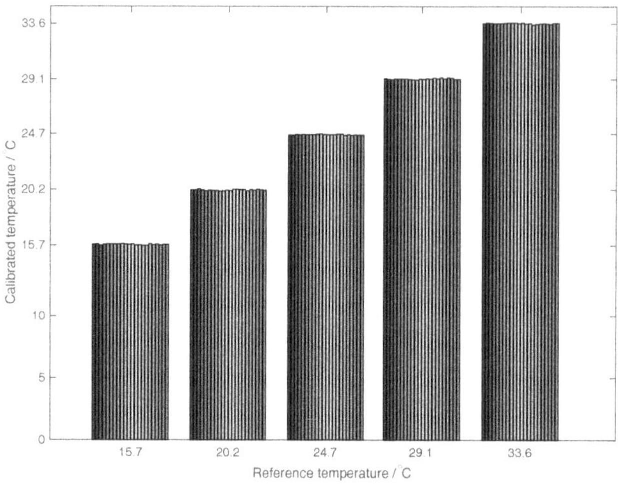

Figure 6.3.: The measured temperatures adjusted using the calibration coefficients. The absolute error is reduced to $< 0.2K$.

the thermocouples to the cylinder is a special compound with high thermal conductivity and low electrical conductivity. The thermocouples were first coated with a small layer of this compound. After the compound hardened, the thermocouples were placed onto the cylinder. The coating does have an influence on the reaction time of the temperature measurement. To determine this time I took a thermocouple with coating and one without. A thermostat was heated above room temperature. The thermocouples were dipped into the hot water at the same time. The measured temperatures were logged on a PC and after evaluation of the data it was possible to determine the time at which the temperatures increased. The difference between these times is the delay of the reaction time caused by the coating. The coating increases the reaction time by about $200ms$.

One of the challenges when evaluating the temperature data was the short timeframe of the μg phase. When entering the μg first the flow pattern changes. This takes at least $5s$. Then, as a result of this change, the heat transfer will also change. This means that there are only $15s$ or less, where the changed heat transfer is in effect. This is long enough for the thermocouples to adapt. But the aluminum outer cylinder has a high heat capacity. This slows down the rate at which the changed temperature is measured. The observed changes were usually lower than $0.5K$ which is nearly in the same range as the measurement errors.

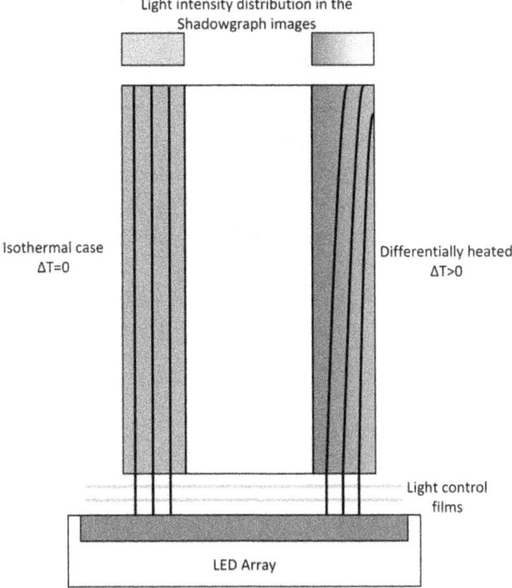

Figure 6.4.: Schematic of the Shadowgraph method. Light is emitted by the LED array and rectified by the light control films to achieve nearly telecentric light. When no temperature gradient is applied, the light will travel undisturbed through the fluid and the camera will see a homogeneous image. When a temperature gradient is applied, then the light will be refracted towards the colder parts of the fluid. The Shadowgraph image shows a gradient in the light intensity.

6.2. Shadowgraph method

The shadowgraph method provides an easy to use and non-invasive method to qualitatively visualize the changes of the refractive index within a transparent liquid or gas (Schöpf *et al.* 1996). The refractive index depends on the density of the fluid. The density on the other hand depends on the temperature and pressure. Since non-compressible fluids in a non-pressured system are used, I can assume the pressure to be constant. Thus, all changes detected by this method are caused by changes in the temperature.

The gap is illuminated from the bottom and the image is acquired via a surface mirror from the top of the cell (Figure 5.5). This image is converted to a grayscale image. Since the light source is red, this is done by extracting the red channel from the images. Before setting the experiment parameters a reference measurement is performed. This means

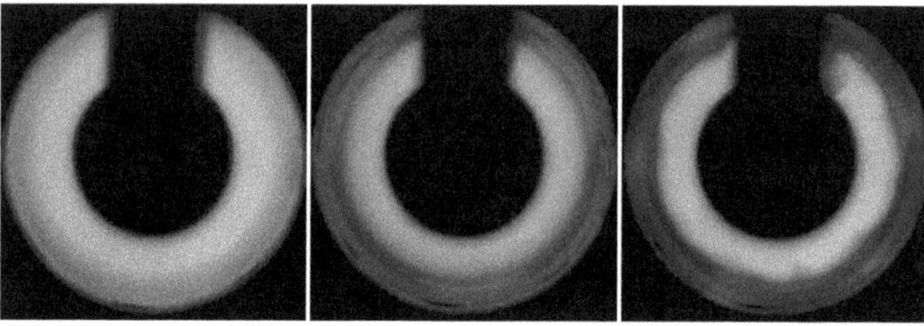

Figure 6.5.: Actual measurements with 3 different parameters. Left: No temperature gradient nor voltage is applied. Middle: A temperature gradient, but no voltage, is applied. The region near the outer cylinder becomes darker, because the light beams are refracted towards the cold outer cylinder and do not reach the upper end of the cell. Right: A temperature gradient and voltage is applied. The distribution of the light intensity becomes less homogeneous and dents appear (e.g. bottom middle).

that no temperature gradient or voltage is applied. Then the experimental images with applied temperature gradient and electric field are acquired. It is not required to use a reference measurement, but using it enhances the contrast and the influence of the electric field is easier to see. The final image is calculated by $I_r = \frac{I_{measured}}{I_{reference}}$, where I denotes the light intensity given by the gray level of the image. This gives the relative change I_r of the light intensity. The general context is that the respective area is warmer than in the reference when $I_r > 1$ and colder when $I_r < 1$. However, this is not always true, especially near the boundaries. I tried to make a correlation between the change of light intensity and change of temperature, however this did not work very well. Using many laboratory experiments I was able to determine the temperature with a precision of $\pm 2K$ when applying the algorithm to other laboratory experiments done under the same condition. When I applied it to data obtained from the PFC the temperatures could be off by $10K$.

Some images are shown in Fig. 6.5 as example of how the untreated experimental images look like. the light intensity inside the gap is nearly homogeneous when no external forces are applied. When the temperature gradient is applied, the light beams are refracted towards the colder medium. The result is that the area near the inner cylinder becomes darker, because the light is refracted towards the outer cylinder. The beams cannot reach the top of the cell anymore since they are bent towards the cell wall. The reflections caused by this are minimized, because the outer cylinder is anodized matte black. When also a voltage is applied the light distribution becomes less homogeneous and certain structures seem to form in the middle of the gap. This cannot be seen very well in the original images, but can be accentuated by the aforementioned principle.

Antoine Meyer wrote several Matlab scripts to calculate the refractions of the lightbeams inside the differentially heated fluid (Meyer 2017). The goal is to find the parameters where this methods works the best and to find the qualitative images for the expected flow pattern. The number of azimuthal and axial modes, height of the cell, and temperature gradient have to be given to the function. This function uses a simple analytical approach to compute the 3D temperature distribution inside the gap. A telecentric light source is assumed at the bottom of the gap. The light path through the gap is calculated based on the calculated temperature profile and the temperature-dependent changes of the refractive index. The image produced by this function is the relative light intensity in the most upper layer of the calculated profile.

The results for a toroidal flow pattern in a 100mm cell are shown in Fig. 6.6. The upper diagram shows the trajectories of the light beams in the $z - \phi$ plane. The lower image shows the calculated relative light intensity at the top of the cell. In a toroidal flow pattern are counter-rotating helices along the height of the gap. Alternating hot fluid is moved from the inner to the outer cylinder and cold fluid is moved the other way around. This is shown in Fig. 6.6 top by the alternating red and blue areas. By integrating the temperature over the height of the gap, the average temperature is determined. And along with that the average change of the refractive index is calculated. This is shown in Fig. 6.6 bottom. The relative light intensity is also used to plot the experimental images, which makes it convenient to compare the simulations to the experiment. Because of the alternating hot and cool vortices, the light traveling from the bottom to the top of the gap is refracted in a zig-zag pattern. Integrating over the height will negate all effects by the inhomogeneous temperature distribution in the gap and the resulting Shadowgraph image shows a homogeneous light distribution.

The columnar flow pattern produces a very distinct pattern in the Shadowgraph image (Fig. 6.7. The alternating vortices are located along the azimuthal direction in the gap. Integrating over the height results in a strong amplification of the refraction of the light beams. This leads to bright spots in the Shadowgraph image, which are caused by the cold jets, where the fluid is moved from the outer to the inner cylinder.

These two patterns are the extreme cases. In an experiment it is possible to see both pattern at the same time. The columnar pattern can also become a helicoidal pattern when the columns are slanted.

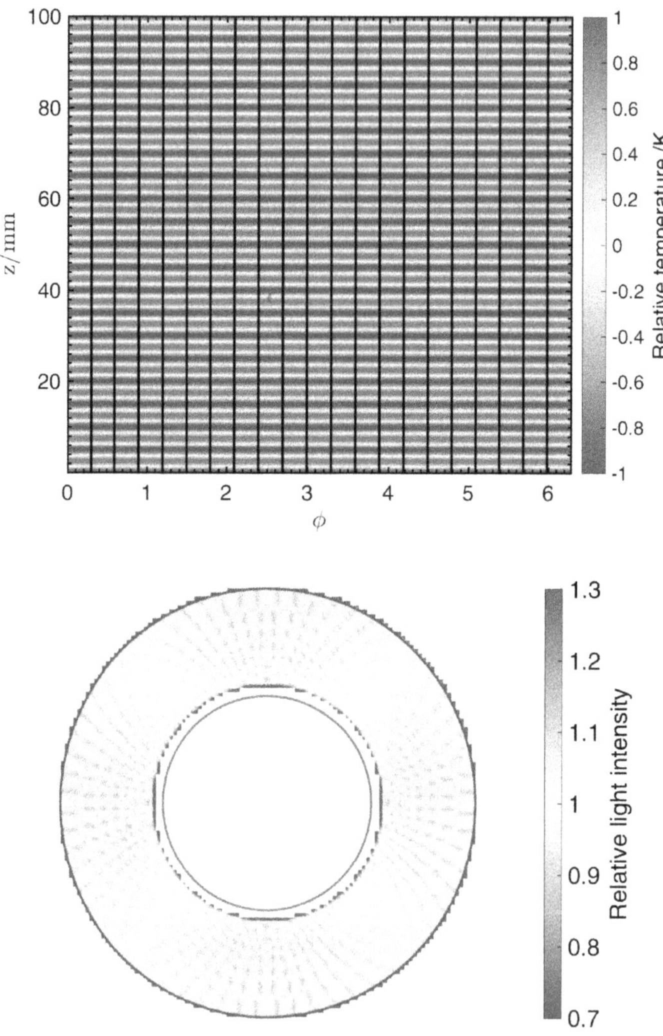

Figure 6.6.: Computed trajectories of the light (top) and the relative light intensity (bottom) of a toroidal flow pattern in the gap. Because of the zig-zag pattern of the light beams, the resulting image is nearly homogeneous (A. Meyer, to be published).

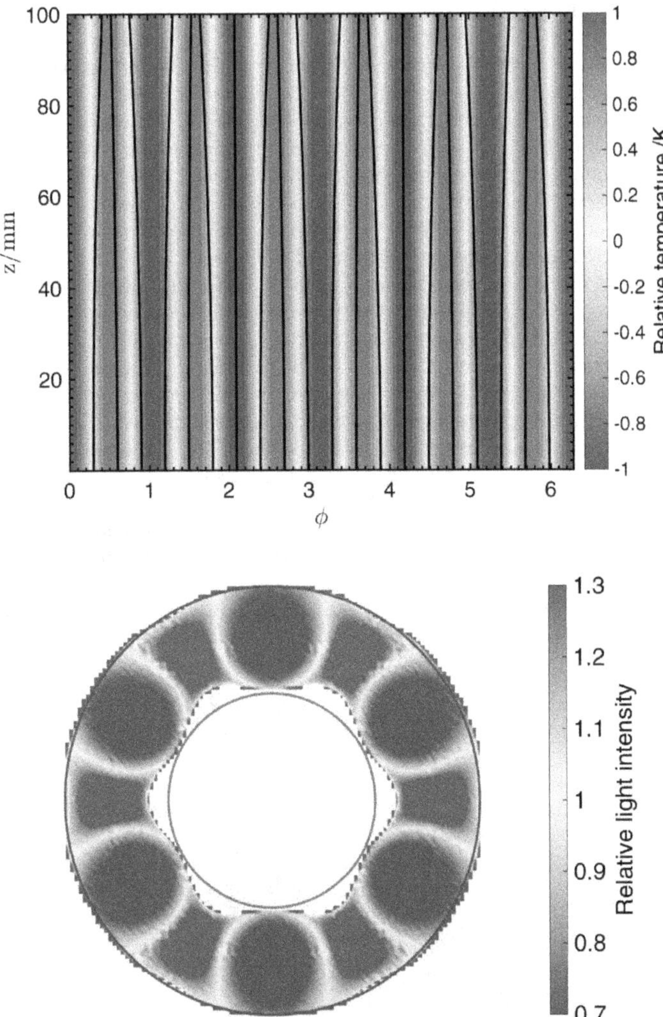

Figure 6.7.: Computed trajectories of the light (top) and the relative light intensity (bottom) of a columnar flow pattern in the gap. The light beams converge at the colder regions and diverge at the warmer regions. The jets of hot fluid going from the inner to the outer cylinder are visible (A. Meyer, to be published).

6.3. Synthetic Schlieren technique

The Shadowgraph method produces very good images, which allow for an easy determination of axisymmetric and non-axisymmetric patterns for the $100mm$ cell. However, for the $30mm$ cell it does not work as well as for the $100mm$ cell. The light path for the columnar pattern (Fig. 6.7) shows points of convergence and divergence at $100mm$ height, but the difference between the starting condition and $30mm$ height is barely noticeable. Because of the lower height, the light is refracted to a lower degree and the effect seen by the cameras is too low to be evaluated. To get a similar result for the $30mm$ cell, a different measurement technique had to be found.

The Synthetic Schlieren method (Dalziel *et al.* 2000, Raffel 2015, Richard *et al.* 2001) uses a similar measurement setup as the Shadowgraph method. In addition to the setup shown in the previous section, there is a mask introduced between the light source and the fluid containmnet. The camera is focused on the pattern on the mask instead of the top of the cell. Several patterns for the mask are described in the literature. For example line or grid based patterns can be used. A matching algorithm is used to determine the distortion of the experimental image to a reference image. It is also possible to use a dot pattern and apply a particle tracking algorithm to calculate the exact displacement of the dots. The disadvantages of this methods are that the spatial resolution is limited by the pattern and an evaluation is not possible when the distortions are too big. The background oriented schlieren (BOS) method as described by Raffel 2015 and Richard *et al.* 2001 does not have this disadvantages. It uses a mask with randomly distributed dots (e.g. 6.8), which increases the spatial resolution, and is evaluated using a cross-correlation similar to the PIV method.

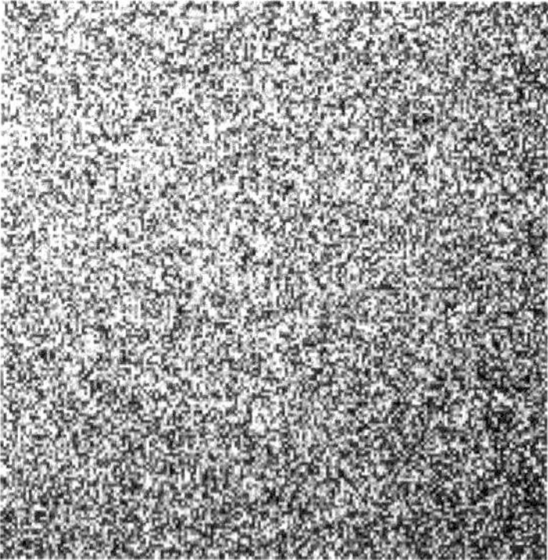

Figure 6.8.: The random dot pattern used as mask for the Synthetic Schlieren method.

The dots introduce a sharp distinction between a dark and bright spot on the image sensor of the camera. There is no gradient like in the Shadowgraph method. This makes this method more sensitive to small changes in the light intensity. The original Schlieren technique introduces a knife edge in the focus point of the light beams between the fluid and the camera. All light beams which are refracted beyond a certain degree are blocked by the edge and do not reach the camera anymore.

The evaluation of this images is done with a cross-correlation. A snapshot of an undistorted image is acquired as reference. This is done when the temperature gradient, but no voltage, is applied. It is also possible to use a reference image without temperature gradient and no voltage like for the Shadowgraph method, but the blurring near the inner cylinder makes the results less reliable. Some experimental snapshots are given in Fig. 6.9 to illustrate these changes. The cross-correlation is performed by using the PIVlab toolbox for matlab (Thielicke *et al.* 2014). This toolbox includes a direct cross-correlation (DCC) and a Fast Fourier Transform based (FFT) algorithm. Both methods divide the image in several interrogation windows and matches the areas in the experimental images against the reference. Tests have shown that the FFT approach produces more reliable results.

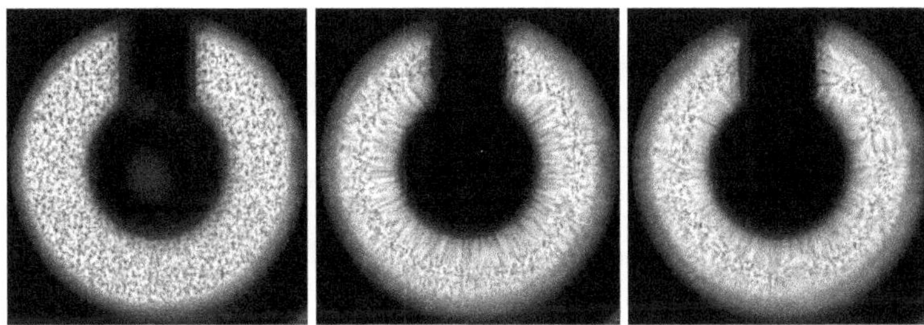

Figure 6.9.: Experimental snapshots of the Synthetic Schlieren setup. Left: No temperature gradient, nor voltage is applied. Middle: A temperature gradient is applied. The area near the inner cylinder is blurred and the dots are stretched in radial direction. Right: A temperature gradient and voltage is applied. In addition to the radial blur by the temperature gradient, there is now also a slight azimuthal blur.

In a first step of the evaluation process the images are cropped to the region of interest and converted to a grayscale image. Then the positions with the reference images have to be defined. The experimental runs during the PFC are planned in a way that the reference is acquired in a $1g$ phase when the temperature gradient, but no voltage, is applied. The cross-correlation is performed with two passes, one with 32×32 pixels window size and one with 16×16 pixels. The image is deformed using the **spline* option and the subpixel finder is the 3-point Gauss algorithm. The resulting vectorfield describes the displacement of the pattern in pixels. It is known that the gap is 5mm wide. The number of pixels in the gap are acquired from the experimental images. This data is used to calculate the calibration factor $c = 5mm/no.ofpixels$. As a result the displacement can be given in mm. The resulting data is presented as false color representation of the magnitude of the vectors. In addition the vectors are plotted as well.

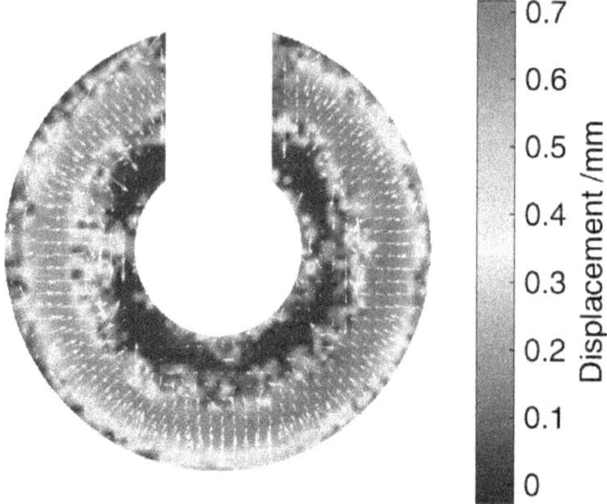

Figure 6.10.: The cross-correlation is done between a reference without temperature gradient nor voltage and the case where the temperature gradient is applied. The magnitude of the displacement is nearly homogeneous and the vectors point inwards. This is identical to the Shadowgraph images, where the outer parts of the gap become darker, when the temperature gradient is increased.

The images in Fig. 6.9 are correlated in several ways to show the difference. The resulting vectorfield is homogeneous with inward facing vectors, when only a temperature gradient is applied (Fig. 6.10) and no external forces are present in the reference. This is expected, as the temperature gradient is oriented in radial direction. When an electric field and temperature gradient is applied, then the choice of the reference image can change the result. When the unheated state is chosen as reference (Fig. 6.11 top), then the columnar structures look similar to the structures obtained from the Shadowgraph method. In the gap are circular structures at which the vectors converge. Unfortunately this only works for $\Delta T < 5K$. At higher gradients, it is not possible to see this structures anymore. This is due to the blur caused by heating the system. The higher the ΔT, the further the blur extends from the inner cylinder inside the gap. This reduces the quality of this images. Using the heated state as reference (Fig. 6.11 bottom) changes the resulting vectorfield. There are now areas near the outer cylinder, where the vectors converge. These can also be related to the columnar structures. But now the vectors point to the cold fluid on the outer cylinder instead of the columns. This approach also works at higher ΔT, since the structures of interest are located near the outer cylinder and are affected less by the blur caused by the temperature.

The Synthetic Schlieren method is overall not easy to use with our experimental setup, since it was developed for very thin gas layers and not liquid layers of $30mm$. But, since the Shadowgraph method did not produce any results in the laboratory pre-experiments I performed, it is used as it allows to determine the azimuthal mode number. The quality of the images is not equal in all parts of the gap and differs highly depending on the used pattern.

To increase the quality of the resulting vector fields, I tried to use different patterns on the mask. Decreasing the size of the randomly placed dots and increasing the printing resolution does, in theory, increase the quality of the matching algorithm, because the interrogation window size of the cross-correlation can be decreased. However, the dots were too fine and the blur of the dots was so strong at higher ΔT, that they were not recognizable on the acquired images. The pixels on the sensor in the camera are too big for this application. Therefore, they need to have a certain minimum size, so that the camera is still able to resolve the dots. The used pattern is a compromise between achievable resolution and wide range of experimental parameters at which this can be used.

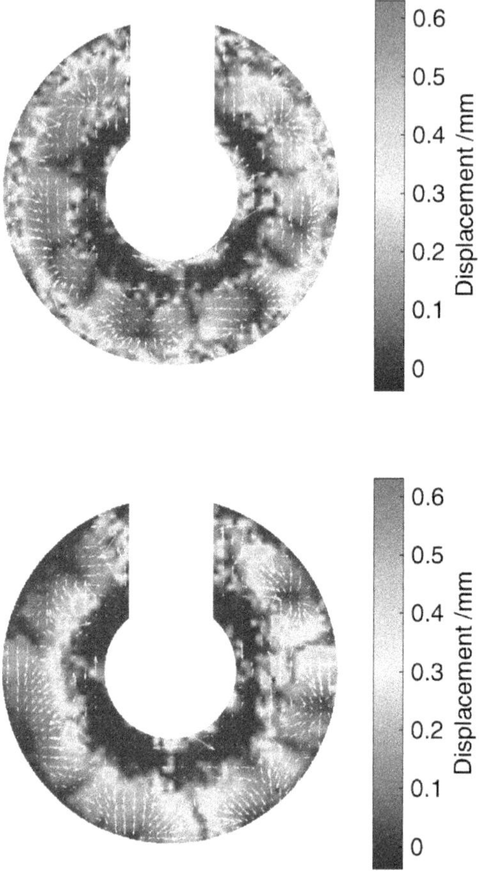

Figure 6.11.: Top: The cross-correlation is done between a reference without temperature gradient nor voltage and the case where the temperature gradient and voltage is applied. Bottom: The cross-correlation is done with a reference with temperature gradient, but no voltage. The vectorfield suggests a columnar-like structure with mode $n = 7$.

6.4. PIV technique

The Particle Imaging Velocitmetry (PIV) is a way to visualize a flow using tracer particles. A schematic of the used setup is shown in Fig. 5.6. A line laser module (see chapter A.1) with continuous laser produces a laser light sheet along the axis of the cylinder. A camera (see table A.1) is mounted perpendicular to the light sheet and acquires images at $10Hz$. The evaluation is done in Matlab with the PIVlab toolbox (Thielicke *et al.* 2014).

The algorithm takes two images in a time series. The first image is divided into interrogation windows with a variable size. Then a cross-correlation is done with the second image to find the position of the interrogation window in the second image. This is repeated for every part of the image. The result is the displacement of each of the windows. Since the time that elapses between the two images is known, it is possible to calculate the velocity from the displacement. This approach makes it possible to calculate a 2D velocity field from videos (Adrian 1991, Raffel *et al.* 2007).

The principle of this process is shown in Fig. 6.12. To remove static noise a simple preprocessing step is performed. The mean over all images of a data series is calculated. This creates an image where the moving particles nearly disappear, but static effects remain. It is possible to see reflections caused by the laser and also the wires, which are used to make electric contact to the cylinder, at the top of the image (Fig. 6.12 A, middle). This background image is subtracted from the experimental images to remove the static noise, which reduces the error in the results of the cross-correlation. The cross-correlation (given by $\star$) is done between two corrected experimental images, which are recorded at a different time (Fig. 6.12 B). The first image is divided into interrogation windows of 32×32 pixels (highlighted in left image). In the second image a 64×64 pixel window is highlighted. The center point of these images is the same. The result of the cross-correlation is a 2D matrix. The 32×32 pixel window is moved pixel-wise over the 64×64 pixel window. For each step a magnitude is calculated which describes how good the data matches. The higher this magnitude the better the match. The 32×32 pixel window has moved to the position inside the 64×64 pixel window where the magnitude is the highest (Fig. 6.12 B, right). In this example the window moved 0 pixel on the x-axis and 9 pixel upward on the y-axis. The pixel movement can be converted into meters using a calibration. The gap width is $d = 5mm$ and the image width is $w = 96$ pixel. The ratio is $5mm/96\text{pixel} = 0.052mm/\text{pixel}$. Thus, the interrogation window moved upwards by $9\text{pixel} * 0.052mm/\text{pixel} = 0.47mm$. This is the displacement given as result for the Synthetic Schlieren images. The time step between the two images, which is $t = 0.5s$, is included to calculate the velocity $u = 0.47mm/0.5s = 0.94mm/s$. This process is repeated for all interrogation windows to create the velocity field of the whole image.

The used tracer particles are Potters Hollow Glass Spheres (HGS) with $\rho = 1.1g/cm^3$, $\varepsilon_r = 4.8$ and $d \approx 10\mu m$. The tracer study was started by N. Dahely (Dahley 2014) and continued by our group with different particles. Tracer with an electrical conductive coating have been ignored, since they will cause flashovers inside the electric field. Polyamide particles, which were used previously, are agglomerating inside the electric field.

They can be used, but only for a short time and need to be mixed after every experimental run. Another negative example is particles made from ceramics, which would not follow the flow but the electric field. However, the HGS work perfectly fine for AK5 and follow the flow also while the voltage is active. The permittivity of the particles is higher than the permittivity of the fluid. I performed some experiments to verify that they move independently from the electric field. The fluid was in an isothermal state and an electric field with $10kV$ peak voltage was applied. The frequency was increased slowly starting from $0.1Hz$. For lower frequencies ($f < 50Hz$) it was visible that the particles moved along the electric field. For higher frequencies this did not happen. I can conclude that the particles are only affected by the dielectrophoretic force at $f = 200Hz$. Further information can be found in chapter 7.5.

The parameters used for the PIV evaluation are dependent on the experimental parameters. For all experiments, a multi pass approach with a window size of 32×32 pixels and 16×16 pixels was used. The gap of the experiment cell is only $5mm$, which is about 90 pixels in the image, depending on the cell type. Lower sizes for the interrogation window are not favorable because then there are not enough particles inside the interrogation windows to ensure a good result from the cross-correlation. The interrogation windows are moved by 50% or 75% during each step. This increases the number of grid points while still keeping a moderate size of the window to minimize errors in the correlation. The sub pixel interpolation was done with a $2 \cdot 3$-point fit. The time step between the images varies depending on the experimental parameters. In μg without external forces, the movement of the particles becomes nearly zero. In this case the time step is $1s$. When high temperature gradient ($> 10K$) or voltages are used, this step can be as low as $0.2s$.

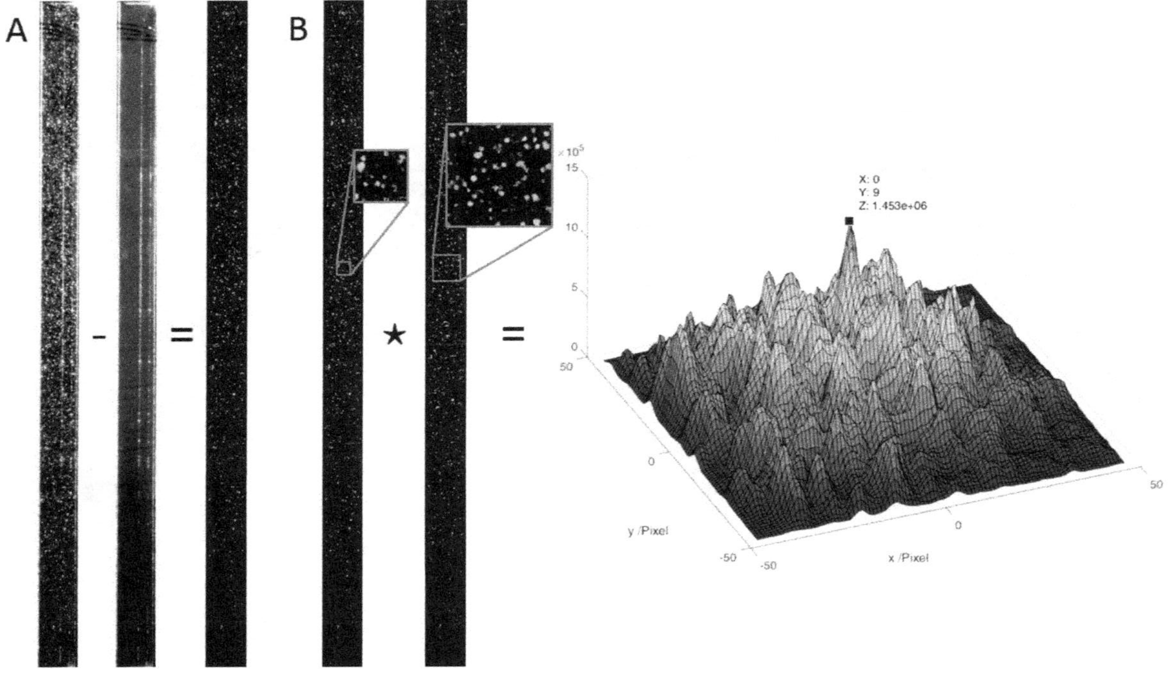

Figure 6.12.: A: In a preprocessing step the average of all images of the data set is calculated. The resulting background image shows static interferences like reflections and is subtracted from the experimental images. B: The experimental images are divided into interrogation windows. The displacement of these windows is calculated using the cross-correlation.

7. Results

During my time as PhD student I took part and collected data in 4 PFC concerning the effects of TEHD forces in a cylindrical annulus. Different fluids with different initial conditions have been tested during the PFC as well as in the laboratory. The used fluids differ in their properties such as permittivity and viscosity.

The chapter is split into several sections to differentiate between the used cell heights. The $30mm$ cell with an aspect ration of $\Gamma = 6$ is governed by boundary effects, while the $100mm$ cells with $\Gamma = 20$ shows less influence of boundary effects. In addition, there are some experiments with different initial conditions, where the voltage is applied at different points of the μg phase. It can be activated at the beginning of the μg phase, which gives an unaffected base state from the $1.8g$ phase. Or it is activated $10s$ into the μg phase, which leads to a conductive state as initial condition. However, this reduces the μg time during which the electric field is active to about $10s$.

7.1. Process of evaluation

The classification of the flow pattern is done by combining evaluated PIV, and Shadowgraph or Synthetic Schlieren images (Jongmanns, Meier, *et al.* 2018). It is possible to determine the axial mode number k by counting the vortices in the PIV images. The azimuthal mode number n is determined by the Shadowgraph or Synthetic Schlieren images respectively. Based on the results from the LSA, the possible flow patterns are given in table 7.1.

Table 7.1.: An overview over the mode numbers and the flow pattern connected to these numbers. The pattern of the base state, when no perturbations are visible, depends on g_z.

	$n = 0$	$n \neq 0$
$k = 0$	Base state $g_z \neq 0$: Unicellular $g_z = 0$: Conductive	Columnar
$k \neq 0$	Toroidal	Helical

Figures 7.1, 7.2 and 7.3 show the different flow pattern based on experimental results. The PIV images are colored using the axial velocity v or the radial velocity u, but the vectorfield always shows both components. In addition a Shadowgraph image or

Synthetic Schlieren image is shown for the 100 mm or $30mm$ cell respectively. Twog_z curves are given, because the different visualization methods were applied on different days. In some cases they were even done in different PFC. Many of the Synthetic Schlieren images for the 30 mm cell were acquired in 2016 while the PIV images were acquired a year later.

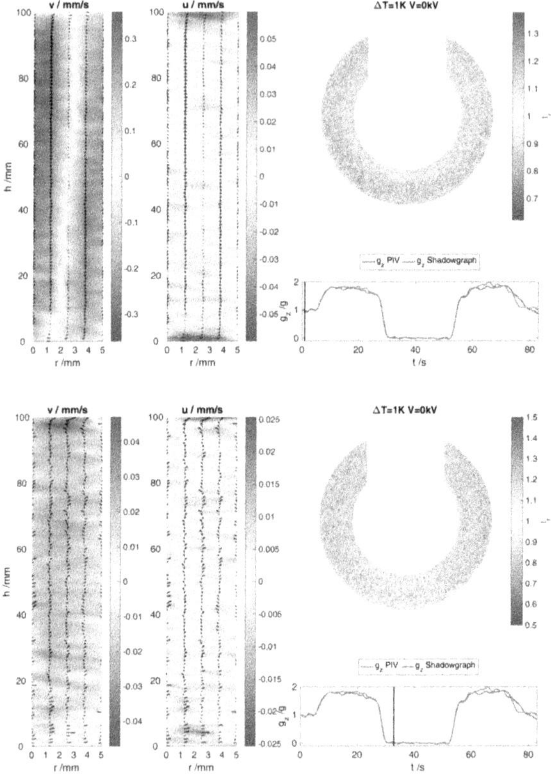

Figure 7.1.: Both possible base pattern with $\Delta T \neq 0$ and $V = 0kV$. In $1g$ conditions in the laboratory (top) a unicellular flow pattern develops. The PIV image shows a single convection cell. The fluid moves upwards at the inner cylinder and downwards at the outer cylinder. In μg (bottom) are no movements of the fluid and a conductive state will appear. In both cases it is not possible to see any structures in the Shadowgraph image, which makes it axisymmetric.

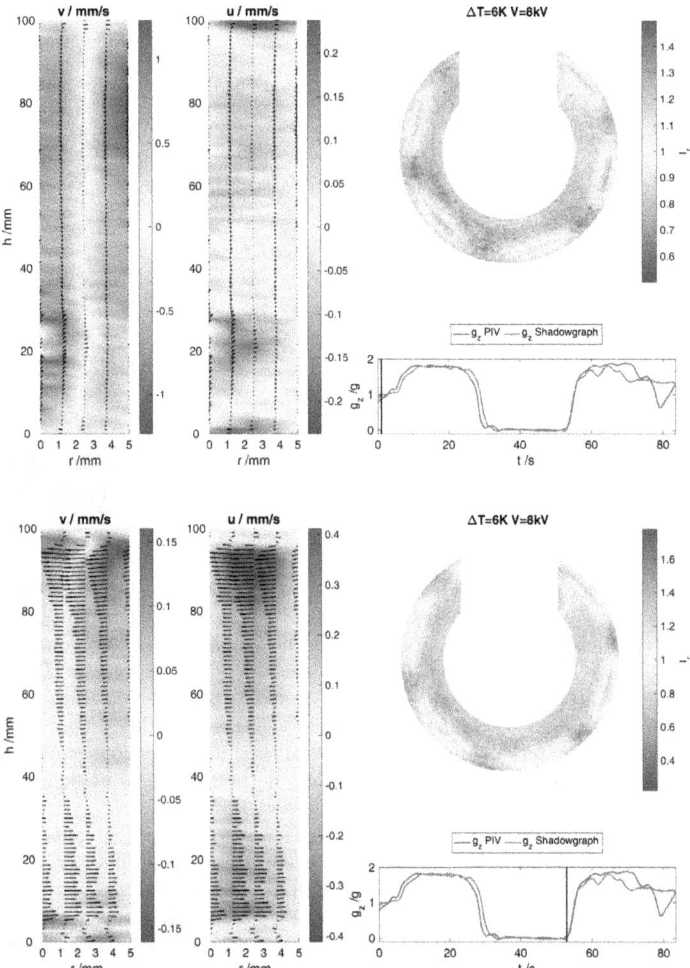

Figure 7.2.: Top: The columnar pattern seen in the $100mm$ cell at $\Delta T = 6K$, $V_p = 9kV$ and $g_z = 1g$. The PIV images shows the onset of an instability in the lower part. The Shadowgraph image shows a structure which is caused by the columns. Bottom: When $g_z = 0g$, then the columns become slanted. This can be seen in the u component of the vectorfield, because it changes the sign. This indicates the border between two counter-rotating columns.

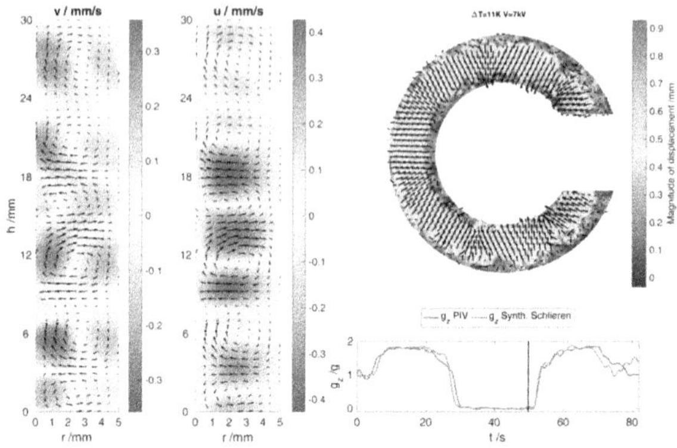

Figure 7.3.: The toroidal pattern seen in the 30mm cell at $\Delta T = 11K$, $V_p = 7kV$ and $g_z = 0g$. The PIV image shows counter-rotating vortices, while the Synthetic Schlieren image shows a slightly disturbed inward pointed vector field.

The base state of the thermal convection, when no voltage is applied, differs depending on the gravity (Fig. 7.1). In the laboratory, the base flow is a buoyancy flow. The gap is heated on one side and the density of the fluid in the gap decreases. Consequently the fluid will raise to the top of the fluid column, where it is pushed to the outer cylinder by the volume flow. There it cools down and the gravity pulls the fluid to the bottom of the column, because the density increases again. Under μg , this cannot happen, because there is no gravitational force. There are no thermal convections and the heat is transported only by conduction. Since there are no perturbations in azimuthal or axial directions, both mode numbers are $n = 0$ and $k = 0$.

There can be two different non-axisymmetric patterns (Fig. 7.2), the columnar and the helical pattern. In both cases it is possible to determine the azimuthal mode number by counting the blue structures in the Shadowgraph images. In these cases $n = 6$. The difference between the columnar and helical pattern is only visible in the vectorfield of the PIV images. While the columnar pattern shows an onset of an instability near the bottom of the cell, I still define $k = 0$, since there is no fully developed mode. The helical pattern shows a gradual change with a changing sign in the u component of the velocity. The change occurs at the border between two counter-rotating columns. While this is also no fully developed mode, I still define this as $k = 1$ to make a differentiation between the columnar and helical pattern.

The toroidal pattern (Fig. 7.3) is an axisymmetric pattern. The Synthetic Schlieren or Shadowgraph images show a mostly homogeneous change without specific structures. However, the vectorfield in the PIV images show several counter-rotating vortices.

Counting the vortices gives an axial mode of $k = 4$.

At some parameters near the transition regions between two different pattern and in the $30mm$ cell, it is not always possible to give a reliable evaluation. The toroidal vortices are promoted by the boundary effects, which have a big effect in the $30mm$ cell. Since the time in the μg phase is limited, it may not be possible to determine the final flow pattern. The conclusions given in this chapter are based around the limitations of the parabolic flight.

The flow pattern are presented as space-time graph of the u component of the PIV vectorfield. The radius is set to the middle of the gap and the graph is plotted over the height and the time of one parabola. The actual g_z level is given below this graph. The μg phase always starts at $t = 30s$. To complement this graph, one single snapshot from the Synthetic Schlieren or Shadowgraph images is given. This is taken at the end of the μg phase, which is $t \approx 51s \ldots 54s$.

The experimental parameters are nondimensionalized to the dimensionless electric potential V_E using equation 3.20 and the dimensionless temperature gradient $\gamma_e = e\Delta T$, where e is the thermal coefficient of permittivity. For AK5 $e = 1.065 \cdot 10^{-3}K^{-1}$. These numbers can be directly compared to the critical values of the LSA done by Yoshikawa (Yoshikawa *et al.* 2013). The LSA also predicts that the flow pattern is always helical in μg , when the critical parameters are applied.

7.2. Flow pattern 30mm cell

First some evaluated experimental images are presented for the $30mm$ cell. The PIV images show a space-time diagram of the u component of the velocity at the middle of the gap. The g_z graph below this diagram is aligned so that the μg phase starts at $t = 30s$. The end is variable and changes every parabola. The Synthetic Schlieren image is taken at the end of the μg phase. Both images can be used to qualify the flow pattern and determine the final flow pattern at the end of the μg phase.

The section is divided into several subsections to differentiate between the starting conditions. The voltage was either active the whole flight, activated when the μg phase started (i.e. $t = 30s$) or $10s$ after the μg phase started (i.e. $t = 40s$).

7.2.1. Voltage continuously active

At $\Delta T = 11K$ and $V_p = 6kV$, an axisymmetric, toroidal pattern starts to develop near the boundaries of the cell (Fig. 7.4). The vortices at the top are stationary, but grow and increase in velocity. The time in the μg phase is too short, so that the flow does not reach a stationary state. At the end of the μg phase, it is a toroidal flow near the boundaries, while there seems to be no change in the flow at $h \approx 20mm$. The homogeneous pattern of the Synthetic Schlieren image taken at the end of the μg phase also suggests an axisymmetric pattern.

Increasing the voltage to $V_p = 7kV$ increases the growth rate of the flow (Fig. 7.5). The development of the toroidal vortices at the top and bottom of the cell are qualitatively similar, but happens at a higher speed. There is also a coherent structure at $h = 12mm \cdots 24mm$, which interrupts the toroidal vortices. The vectorfield in the Synthetic Schlieren image still shows a structure where all vectors point inward. However, some vectors are slightly bend and seem to converge at certain points at the inner cylinder. Overall, this is still regarded as axisymmetric, toroidal pattern. But there seems to be the indication of columnar patterns.

There is a clear transition in the flow pattern when the voltage is increased even further to $V_p = 8kV$. The space-time diagram still shows a toroidal pattern near the boundaries with a discontinuity at $h = 12mm \cdots 24mm$. Due to the increased g_e the growth rates of the perturbations increase as well. The vectorfield of the Synthetic Schlieren image changed completely. The vectors are now pointing in azimuthal direction or to the outer cylinder. This indicates a non-axisymmetric pattern, although the PIV results still show axisymmetric structures. Overall, this is a non-axisymmetric pattern, with axisymmetric boundary effects.

In all cases there is a unicellular pattern in the $1g$ and $1.8g$ phases, although the voltage is active. The boundary effects seem to stabilize this flow pattern. It changes only in the μg phase.

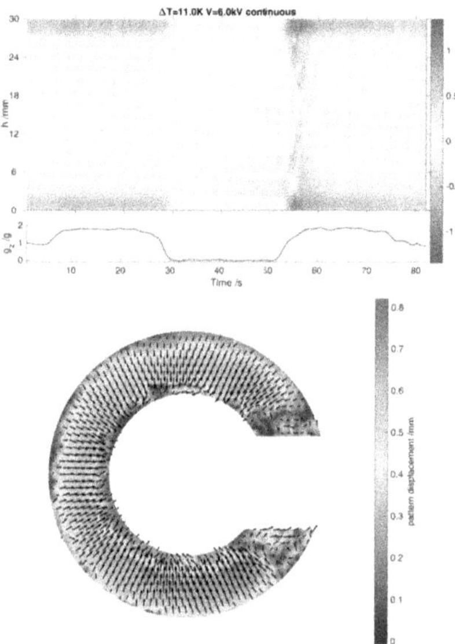

Figure 7.4.: $H = 30mm$, $\Delta T = 11K$, $V_p = 6kV$ continuously active. The space-time diagram shows the onset of a toroidal pattern. The vectorfield of the Synthetic Schlieren image shows an inward directed homogeneous pattern. This supports the existence of a axisymmetric pattern.

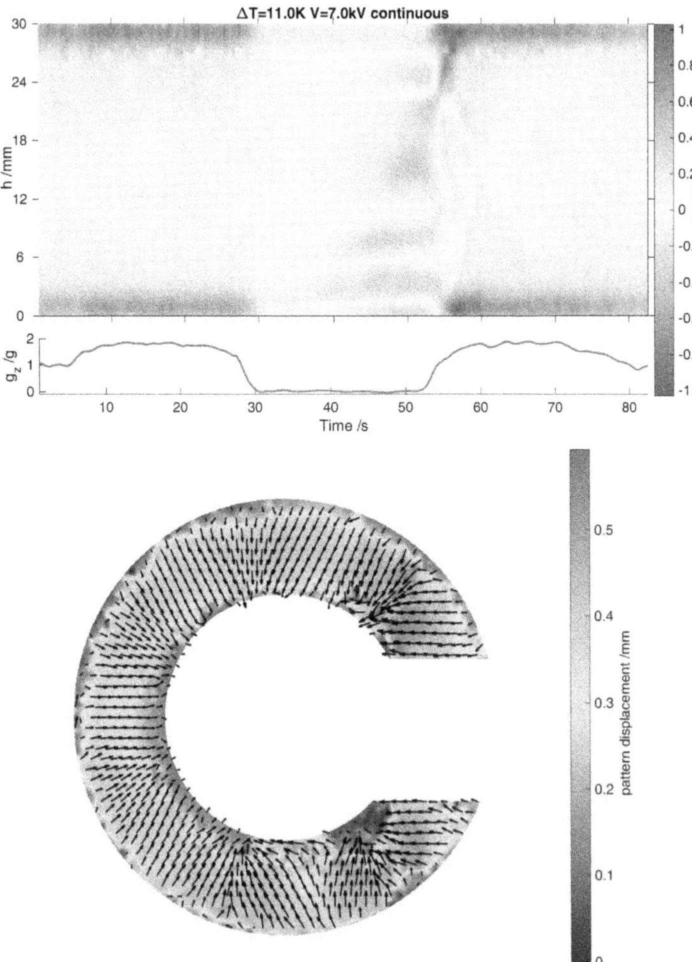

Figure 7.5.: $H = 30mm$, $\Delta T = 11K$, $V_p = 7kV$ continuously active. The space-time diagram shows the onset of a toroidal pattern near the boundaries. At $h = 12mm \ldots 24mm$ is a coherent structure. The vectorfield of the Synthetic Schlieren image shows some areas where the vectors converge. This suggests the onset of a non-axisymmetric, columnar pattern.

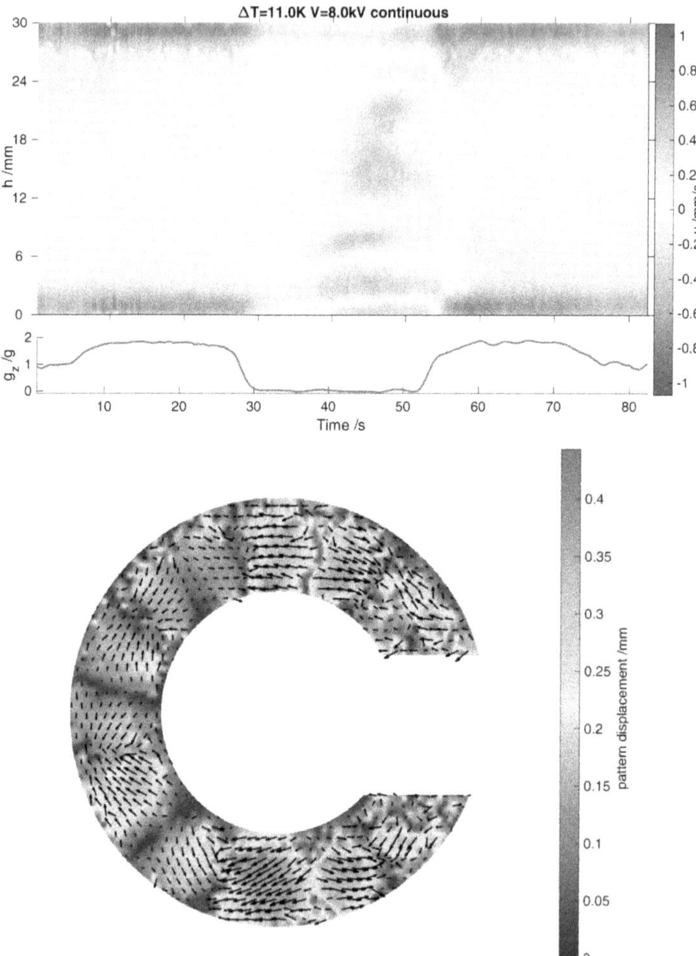

Figure 7.6.: $H = 30mm$, $\Delta T = 11K$, $V_p = 8kV$ continuously active. The space-time diagram shows a similar pattern to $V_p = 7kV$. However, the Synthetic Schlieren image shows a very clear non-axisymmetric pattern. In contrast to the $V_p = 7kV$ parameter, the flow at $h = 12mm \ldots 24mm$ is now a developed column.

7.2.2. Voltage active during μg phase

The resulting flow structure at $\Delta T = 11K$ and $V_p = 7kV$ (Fig. 7.7) does now show a toroidal structure which spans the whole height of the gap. In contrast to the previous subsection, where the voltage was active continuously, there is no discontinuity in the pattern. However, the corresponding Synthetic Schlieren image indicates the onset of a non-axisymmetric pattern in the lower part of the gap. Again, the limited time in μg prevents the flow to become stationary. The evaluation of this pattern is ambiguous. In the overview of the results it would be labeled as non-axisymmetric pattern, which needs to be confirmed in a future PFC.

When the voltage is increased to $V_p = 8kV$ (Fig. 7.8), the resulting space-time diagram is comparable to the diagram in the previous subsection (Fig. 7.6). Toroidal vortices can be found near the boundaries with a discontinuity at $h = 12mm \cdots 24mm$. The vectorfield in the Synthetic Schlieren image shows a non-homogeneous pattern in the whole gap.

Compared to the parameter set where the voltage is active continuously, it is possible to see a pattern with only toroidal vortices over the whole height. The transition from a non-axisymmetric to an axisymmetric pattern occurs in both cases at $V_p = 8kV$.

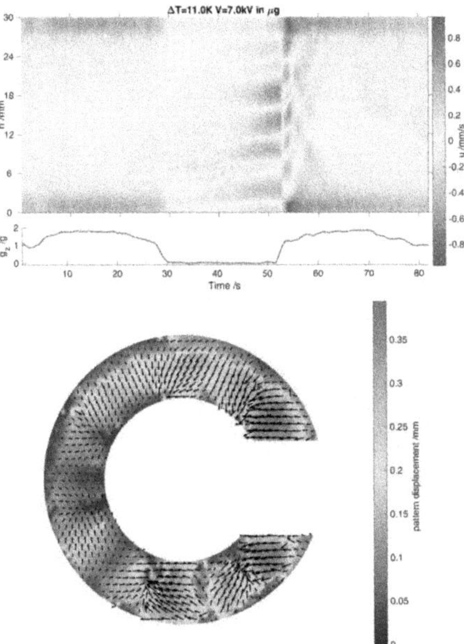

Figure 7.7.: $H = 30mm$, $\Delta T = 11K$, $V_p = 7kV$, which is only active in μg . The space-time diagram shows a toroidal pattern over the whole height of the cell, but the Synthetic Schlieren vectorfield shows the onset of a non-axisymmetric pattern.

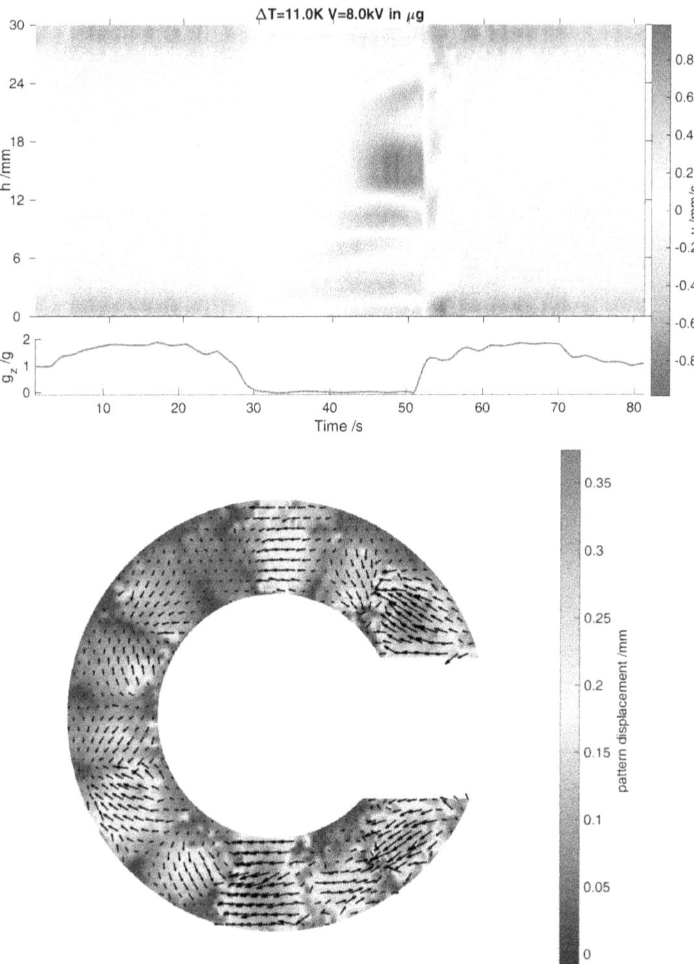

Figure 7.8.: $H = 30mm$, $\Delta T = 11K$, $V_p = 8kV$, which is only active in μg . The space-time diagram shows a toroidal pattern near the boundaries and suggests a non-axisymmetric pattern at $h = 12mm \ldots 24mm$. This non-axisymmetric pattern is supported by the vectorfield of the Synthetic Schlieren image.

7.2.3. Voltage activated $10s$ after μg starts

At $\Delta T = 11K$ and $V_p = 7kV$ (Fig. 7.9), the resulting pattern at the end of the μg phase can be compared to the previous case, where the voltage was activate at the beginning of the μg phase. However, since the voltage is only active about half of the μg phase, the flow is still at the beginning of the transition phase. The space-time diagram shows the onset of a axisymmetric pattern with toroidal vortices. The vectorfield of the Synthetic Schlieren image is homogeneous and the vectors are pointing inwards. Both measurement techniques are suggesting an axisymmetric pattern.

This changes when the voltage is increased to $V_p = 8kV$ (Fig. 7.10). There is still an onset of a toroidal pattern, but there is also a discontinuity in the pattern. In contrast to the previous subsections, this discontinuity is limited to $h = 12mm \cdots 18mm$. The Synthetic Schlieren image suggest the onset of a non-axisymmetric pattern in the lower part of the gap. There is, again, a discrepancy between the results of the measurement techniques. The PIV suggests an axisymmetric pattern and the Synthetic Schlieren image suggests a non-axisymmetric pattern. Overall, this parameter set is near the transition and as treated a non-axisymmetric pattern, which needs to be confirmed. However, since the voltage is activated $10s$ into the μg phase, the time is too limited and it should be done in e.g. a TEXUS flight, where the μg phase is longer.

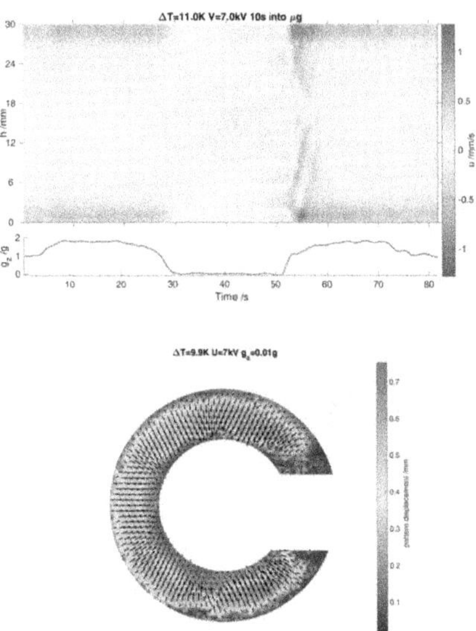

Figure 7.9.: $H = 30mm$, $\Delta T = 11K$, $V_p = 7kV$, which was activated $10s$ after the μg phase started ($t = 40s$). A toroidal pattern starts to develop slowly about $1s$ after the voltage was activated. The Synthetic Schlieren image suggests an axisymmetric pattern.

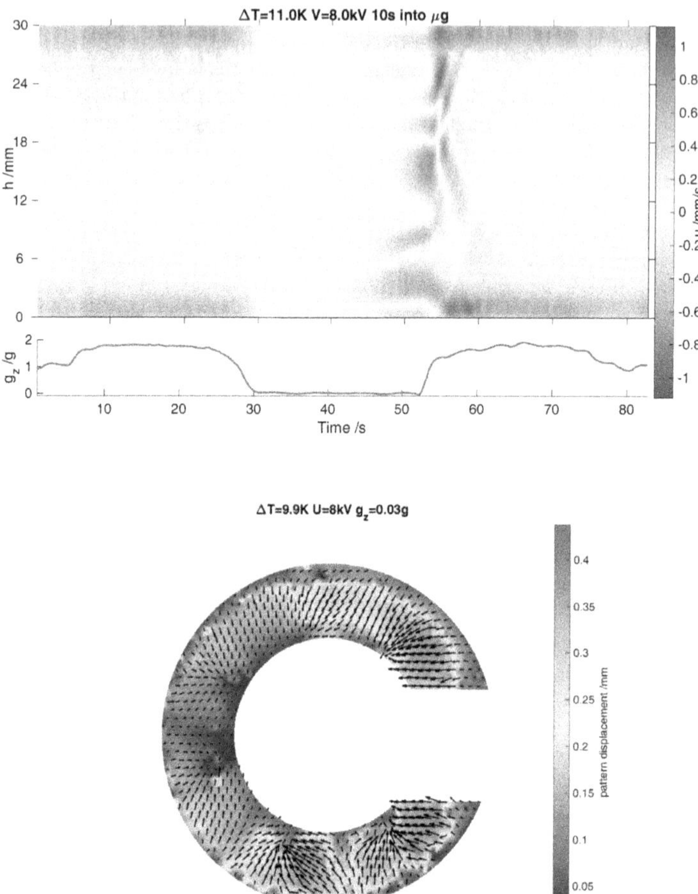

Figure 7.10.: $H = 30mm$, $\Delta T = 11K$, $V_p = 8kV$, which was activated $10s$ after the μg phase started ($t = 40s$). A toroidal pattern starts to develop about $1s$ after the voltage was activated. However, the Synthetic Schlieren image suggests a non-axisymmetric pattern.

7.2.4. Comparison to the LSA

The data of about 75 different parameter combinations where the voltage is applied continuously has been consolidated. The data for $g_z = 1g$ was collected in the $1g$ phases of the parabolic flights and is shown in Fig. 7.11. The data for the μg phase is given in Fig. 7.12. The determination of the flow pattern under μg was done at the end of the μg phase. The dimensionless parameters are used and compared to the LSA in both cases.

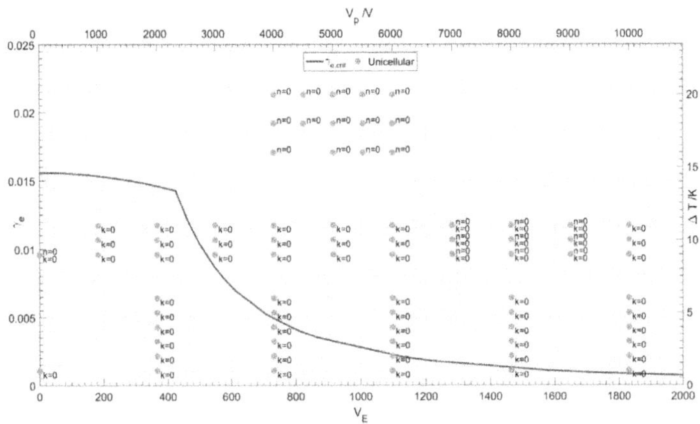

Figure 7.11.: Comparison of the LSA and experiments in $1g$. The blue line shows the critical parameters calculated by the LSA. No disturbances of the unicellular base flow can be seen when $g_z = 1g$.

The $1g$ data always shows a unicellular pattern. Within the parameter range of the experiment (up to $\Delta T = 20K$ and $V_p = 10kV$) I was not able to find any perturbations in the flow. The LSA assumes a cell with infinite height, where the boundary effects can be neglected. However, a small cell with only $30mm$ height is governed by boundary effects. These seem to stabilize the flow in a $1g$ environment.

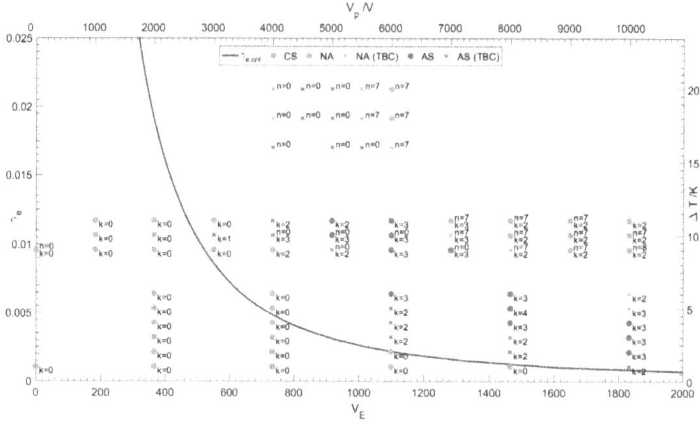

Figure 7.12.: Comparison of the LSA and experiments in μg . The blue line shows the critical parameters calculated by the LSA. CS - conductive state, NA - non-axisymmetric state, AS - axisymmetric state, TBC - to be confirmed.

The observed flow pattern change completely in μg . The base flow, when the experimental parameters are below the critical parameters, is a conductive state. The first observed perturbation is a axisymmetric pattern. This transition fits very well with the LSA for low temperature gradients ($\gamma_e < 0.004$, $\Delta T < 4K$), but the difference becomes larger as the gradient increases. This is not surprising, because the LSA adopts a Boussinesq approximation for the density and a similar approximation for the temperature dependent changes of ε_r. However, the LSA predicts a non-axisymmetric pattern while an axisymmetric pattern is observed in the experiment. This can be explained by the boundary effects. These seem to be toroidal and take up most of the height in the $30mm$ cell. This makes is also difficult to determine where the boundary effects end. The flow becomes a non-axisymmetric pattern after a second set of critical parameters. This pattern is limited to $h = 12mm \cdots 24mm$ in most cases and the other parts of the gap are axisymmetric. Based on the used measurement methods and limitations of the parabolic flight, it is not possible to reliably determine whether they are helical or columnar.

The observed flow pattern do not fit very well to the predictions of the LSA. In μg conditions the transitions occurs at the critical parameters given by the LSA for low temperature gradients. However, the observed pattern is axisymmetric. No perturbations are visible in $1g$ conditions. The boundary effects influence most of the cell height and stabilize the flow. Eventually, the experimental results fit rather well to the LSA, considering the limitations of the LSA and geometry of the cell.

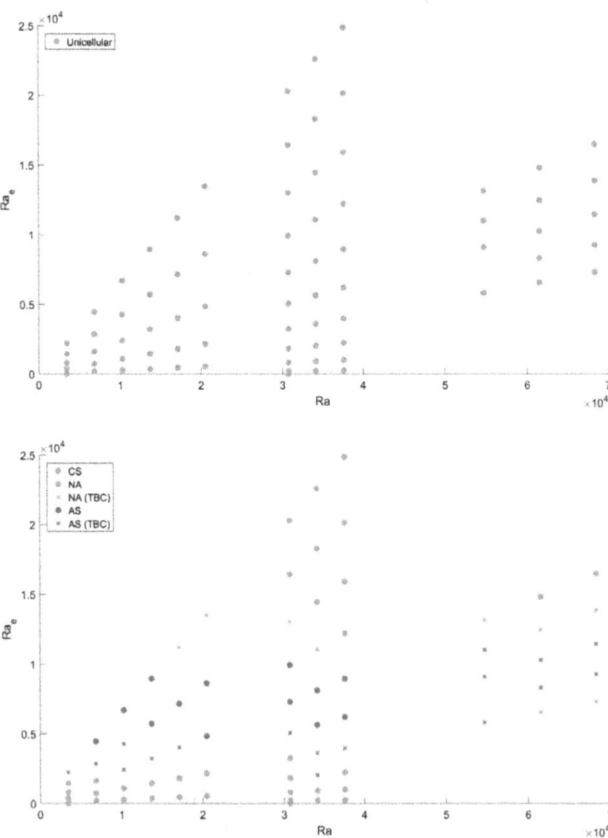

Figure 7.13.: Comparison of the LSA and experiments in $1g$ (top) and μg (bottom) given as $\mathcal{R}a$ and $\mathcal{R}a_e$. The blue line shows the critical parameters calculated by the LSA. CS - conductive state, NA - non-axisymmetric, AS - axisymmetric, TBC - to be confirmed.

7.3. Flow pattern 100mm cell

The PIV results are presented as space-time diagram of the u component of the velocity field. The g_z diagram is also aligned in a way, that the μg phase starts at $t = 30s$. Complementary, I present Shadowgraph images for the $100mm$ cell to qualify the flow pattern in the $r - \phi$ plane. Unfortunately, the Shadowgraph image does not exist for every experimental parameter done with PIV. It is presented whenever possible.

This section is also divided into subsections, depending on the starting condition of the electric field. It is either active continuously, activated at the start of the μg phase or activated $10s$ after the μg phase started. Most Shadowgraph images are only available for the case that the voltage is active continuously.

There are no comparable PIV and Shadowgraph experiments for certain parameters. For example there are no Shadowgraph images for the case where the voltage is activated at the start of the parabola or $10s$ into the μg phase. These parameters were used during my last PFC in 2018 and the corresponding Shadowgraph experiments need to be done in a future PFC. There are also cases where the experiments were performed in different campaigns and not just on different days of the same campaign.

7.3.1. Voltage continuously active

The first parameter set is $\Delta T = 16K$ and $V_p = 7kV$ (Fig. 7.14). The pattern does not show any peculiar features. Based on measurements at $\Delta T = 14K$, I would assume that it is a columnar pattern. However, this needs confirmation. The pattern shifts to a clear columnar pattern in the μg phase. The vectors at the center of the cell point in one direction. On the lower boundary is on toroidal vortex and on the upper boundary is one pair of counter-rotating vortices. The vortices near the boundaries appear about $1s$ into the μg phase and the pattern in the middle starts to appear about $5s$ into the μg phase.

When the voltage is increased to $V_p = 8kV$ (Fig. 7.15), there is also no peculiar pattern in $1g$. Based on measurements at similar parameter, I would, again, assume a columnar patter. In the μg phase, the u component of the velocity field shows a gradient in axial directionwith a change of the sign of u. This is an indicator for a helical pattern. The boundary effects are the same as in the previous case. One toroidal vortex is near the lower boundary and one counter-rotating pair is at the top of the gap.

The following parameters have been performed with both measurement methods, PIV and Shadowgraph. Again, a space-time graph of the u component of the velocity field is shown. The figures show in addition to that PIV snapshots containing the $2D$ velocity field and Shadowgraph images. These snapshots are taken in the $1g$ phase just before the $1.8g$ phase, $5s$ after the μg phase started, and at the end of the μg phase.

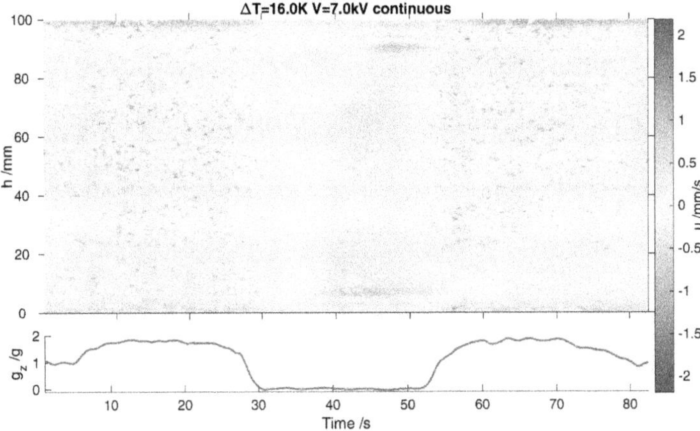

Figure 7.14.: $H = 100mm$, $\Delta T = 16K$, $V_p = 7kV$ continuously active. There are toroidal vortices near the boundaries. In between is a homogeneous pattern which would suggest a columnar pattern.

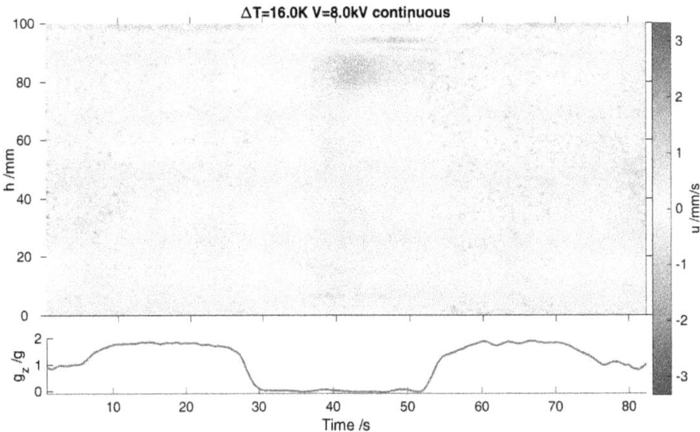

Figure 7.15.: $H = 100mm$, $\Delta T = 16K$, $V_p = 8kV$ continuously active. There are toroidal vortices near the boundaries. In between is a gradient which also changes the sign. This is an indicator for a helical pattern.

The first experimental parameters are $\Delta T = 2K$ and $V_p = 10kV$ (Fig. 7.16). The space-time diagram shows, with the exceptions of boundary effects, a homogeneous pattern, where the vectors are pointing to the outer cylinder. This would be the case in a columnar pattern. The three Shadowgraph images confirm this. In $1g$ as well as μg is a columnar pattern visible. The PIV snapshots show a slightly different pattern in the $1g$ and μg phase. The vectorfield is in μg only influenced by g_e in radial direction. In $1g$ the effect of the natural gravity is visible, since there is an axial component. Finally, the flow pattern is always columnar, but there is still a difference between $1g$ and μg .

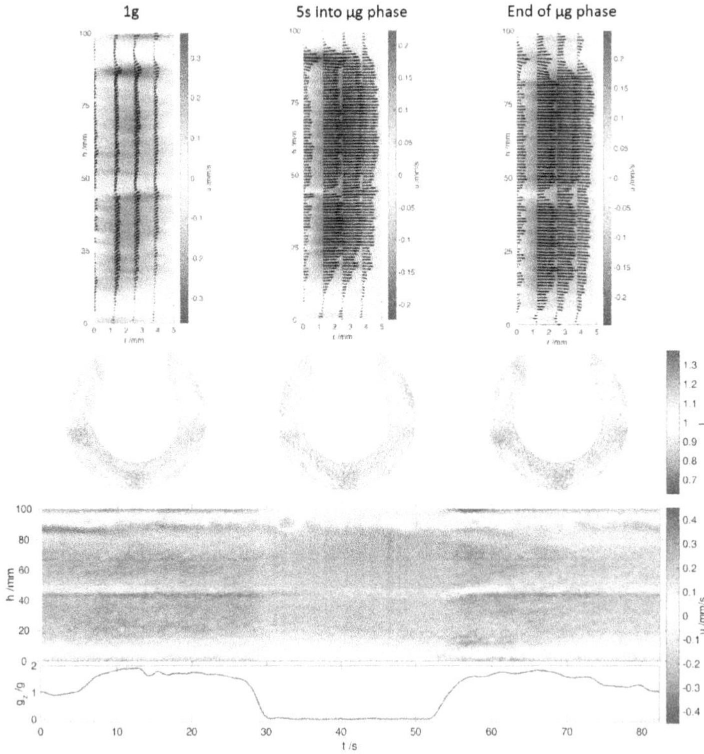

Figure 7.16.: $H = 100mm$, $\Delta T = 2K$, $V_p = 10kV$ continuously active. The Shadowgraph images show a columnar pattern in all three phases and the space-time plot shows a quite homogeneous pattern. The detailed $2D$ vectorfields show slight differences in the $1g$ and μg phase. The vectors have a radial orientation in the μg phase, with the exception of the boundary effects.

The following figures show both results at $\Delta T = 6K$, but different voltages with $V_p = 6kV$ (Fig. 7.17) and $V_p = 8kV$ (Fig. 7.18). At $V_p = 6kV$ the pattern in the $1g$ phase seems to be not influenced by the electric field based on the PIV results. However, the Shadowgraph image shows very feint columnar structures. The voltage seems to be slightly below the threshold needed to establish a columnar pattern during the $1g$ phase. In μg this changes to a helical pattern. This is only visible in the snapshots but not in the space-time diagram, because the velocities are very low. The formation of a helical or columnar pattern is supported by the Shadowgraph images. It is also visible that a re-organisation of the structure happens between the start and end of the μg phase. At $V_p = 6kV$, g_e is barely high enough to cause some disturbances, but it still seems to be a transient state.

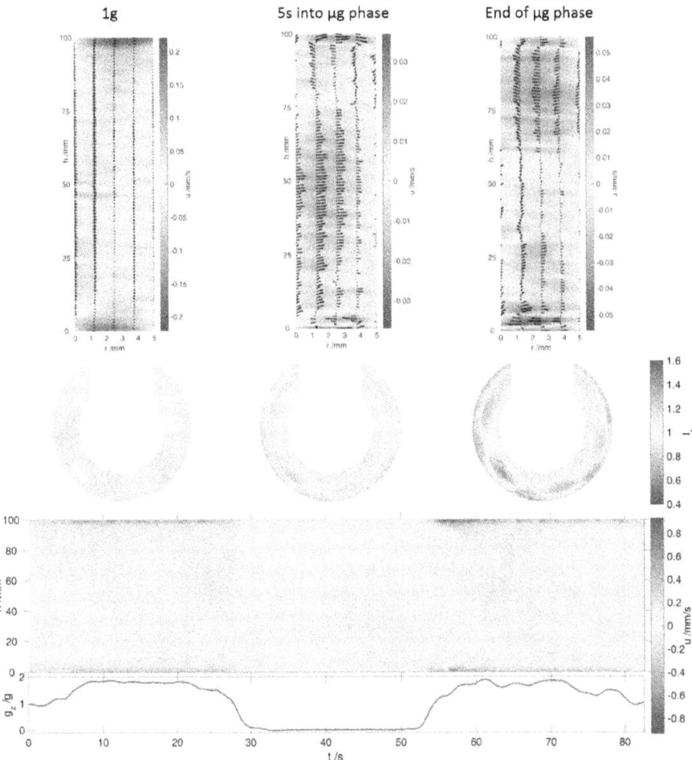

Figure 7.17.: $H = 100mm$, $\Delta T = 6K$, $V_p = 6kV$ continuously active. The space-time diagram suggests an unicellular pattern in the $1g$ and $1.8g$ phases and a conductive state in μg . The Shadowgraph images suggest the onset of a columnar pattern in $1g$ and a further developed columnar pattern in μg .

Keeping the temperature gradient at $\Delta T = 6K$, but increasing the voltage to $V_p = 8kV$ changes the flow pattern. The space-time diagram shows a helical pattern in the $1g$ and μg phase, which changes possibly to a columnar pattern in the second $1.8g$ phase. The Shadowgraph images also support the possibility of a helical or columnar pattern. However, the PIV snapshots add some more detail to the pattern. In $1g$, the u component of the vectors change direction, but only slightly. It could be a helical pattern, but it could also be a vortex at the bottom of the cell, which is created by boundary effects, and a columnar part inside the gap. In μg this changes to a clear helical pattern, where the u component of the vectors are pointing in the opposite direction.

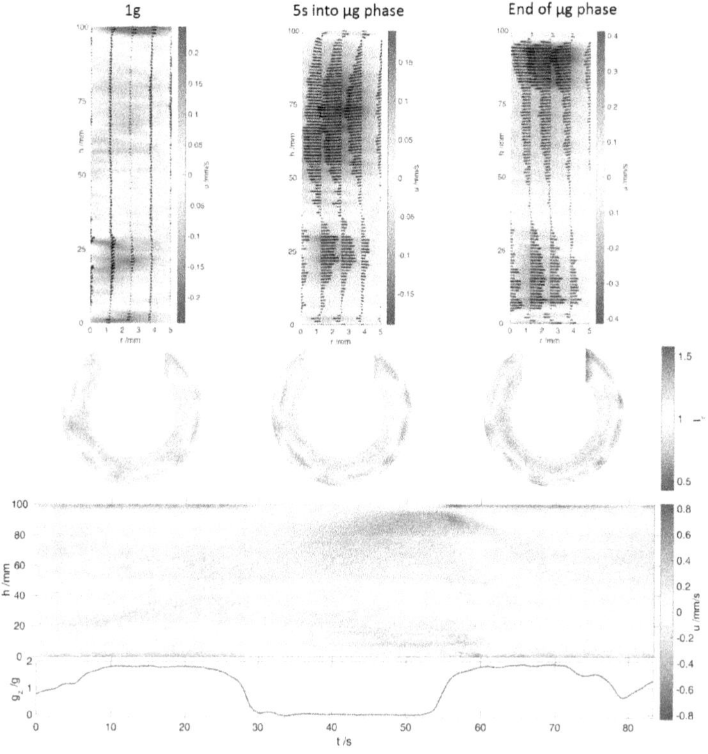

Figure 7.18.: $H = 100mm$, $\Delta T = 6K$, $V_p = 8kV$ continuously active. The Shadowgraph images suggest a columnar or helical structure. The space-time plot shows potentially a columnar pattern in all phases since the sign of the u component changes. The entire vectorfields suggest probably a columnar pattern in $1g$ and helical pattern in μg .

The following figure shows some results at $\Delta T = 10K$ and $V_p = 7kV$ (Fig. 7.19). The snapshots are this time given for the $1g$, $1.8g$, and the μg phase. The space-time plot shows a helical pattern at the end of the μg phase. However, in $1g$ and $1.8g$ it seems to be similar and could be helical or columnar with a large influence of boundary effects. The Shadowgraph images can be interpreted in a similar way. The pattern in $1g$ could be columnar or helical. The structures caused by a columnar-like flow pattern disappear in $1.8g$. The pattern from the $1g$ phase re-appears in the μg phase at higher intensity.

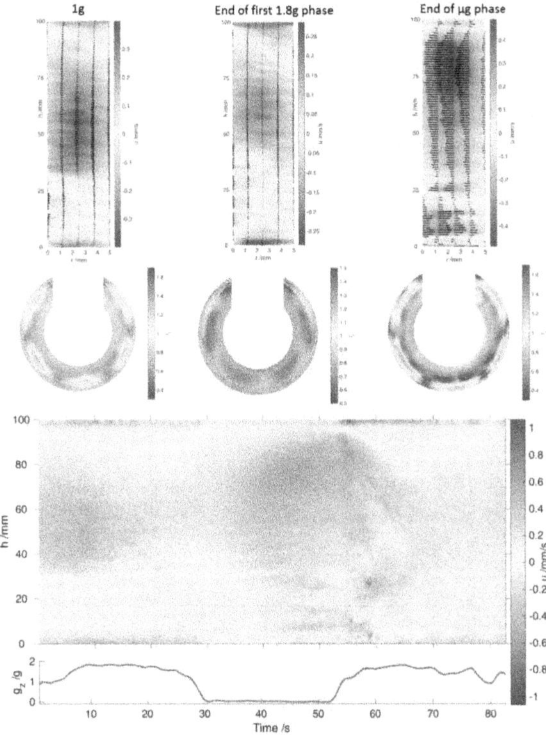

Figure 7.19.: $H = 100mm$, $\Delta T = 10K$, $V_p = 7kV$ continuously active. Based on the space-time plot there could be a columnar or helical pattern in all phases. The Shadowgraph images support this for the $1g$ and μg phases, but the columns seem to disappear in $1.8g$.

The PIV snapshots for the $1g$ and $1.8g$ phase are very similar. However, in the $1.8g$ phase the velocities have a larger axial component than in the $1g$ phase. In the μg phase the axial component of the velocity is nearly zero, with the exception of boundary effects. This parameter set shows a columnar, maybe slightly helical, pattern in the $1g$ phase, a unicellular base flow in the $1.8g$ phase and a helical pattern in the μg phase.

7.3.2. Voltage active during μg phase

Activating the voltage only in the μg phase changes the toroidal patterns at the boundaries. At $\Delta T = 16K$ and $V_p = 7kV$ (Fig. 7.20) there is still only one toroidal vortex at the bottom during the μg phase. At the top of the cell are now several pairs of counter-rotating vortices.

When the voltage is increased to $V_p = 8kV$ (Fig. 7.21), one pair of counter-rotating vortices is at the top of the cell and several pairs are at the bottom of the cell. This is the opposite of what can be seen at $V_p = 7kV$.

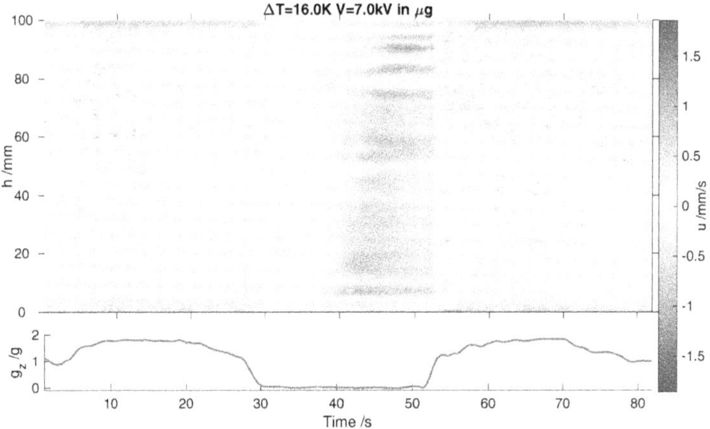

Figure 7.20.: $H = 100mm$, $\Delta T = 16K$, $V_p = 7kV$, which is only active in μg . The main structure in the gap seems to be columnar. There is only one toroidal vortex at the bottom, but several at the top.

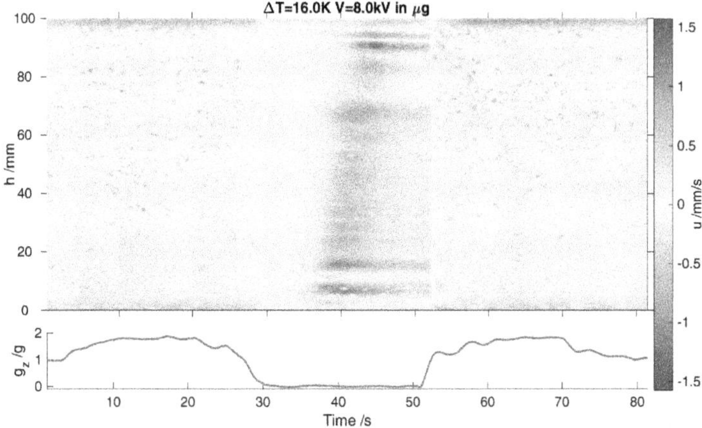

Figure 7.21.: $H = 100mm$, $\Delta T = 16K$, $V_p = 8kV$, which is only active in μg . The main structure in the gap seems to be columnar. There is only one pair of toroidal vortex at the top, but several at the bottom.

7.3.3. Voltage activated $10s$ after μg starts

When the voltage is activated $10s$ into the μg phase, the pattern changes again. In both cases $V_p = 7kV$ (Fig. 7.22) and $V_p = 8kV$ (Fig. 7.23), there is the same pattern. At the upper boundary is one pair of toirodal vortices, while only one vortex is at the bottom. In addition to these boundary effects, one pair of toroidal vortices develops at about $h = 20mm$.

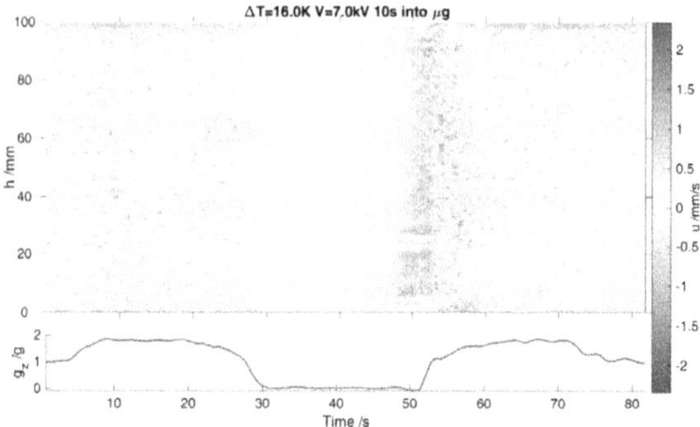

Figure 7.22.: $H = 100mm$, $\Delta T = 16K$, $V_p = 7kV$, which was activated $10s$ after the μg phase started ($t = 40s$). Excluding the boundary effects, only one pair of toroidal vortices is visible at $h = 20mm$.

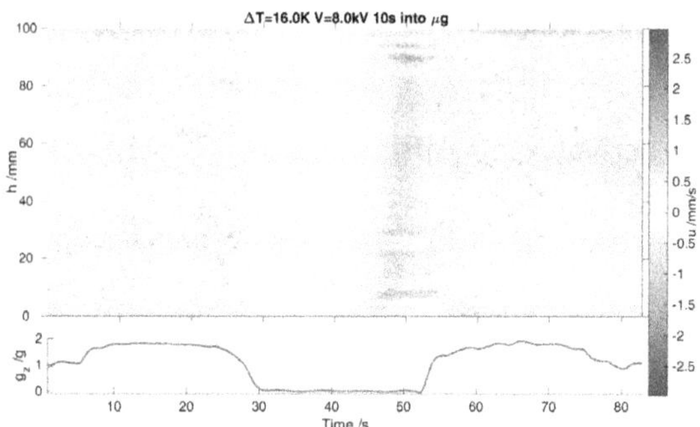

Figure 7.23.: $H = 100mm$, $\Delta T = 16K$, $V_p = 8kV$, which was activated $10s$ after the μg phase started ($t = 40s$). Excluding the boundary effects, only one pair of toroidal vortices is visible at $h = 20mm$.

7.3.4. Comparison to the LSA

About 70 different parameter combinations with continuously active voltage are compared. The $1g$ data is taken from the 1 minute breaks between the parabolas of the parabolic flight (Fig. 7.24). The flow pattern for μg was determined at the end of the μg phase (Fig. 7.25). The experimental parameters are non-dimensionalized and compared to the LSA. When the parameters are above the critical parameters, it is predicted to obtain a helical flow pattern.

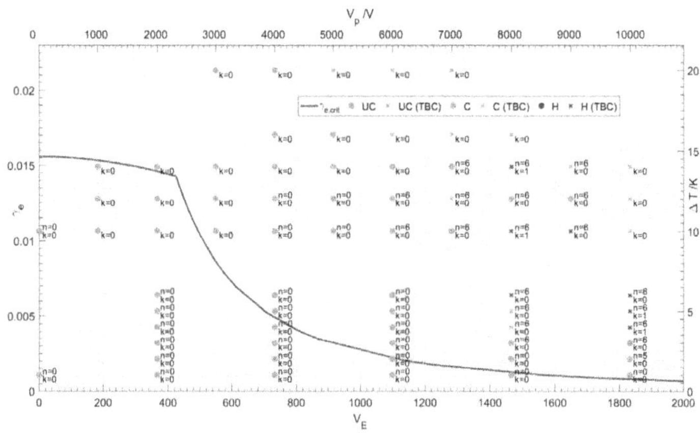

Figure 7.24.: Comparison of the LSA and experiments in μg . The blue line shows the critical parameters calculated by the LSA. UC - unicellular pattern, C - columnar pattern, H - helical pattern, TBC - to be confirmed.

The critical parameters of the LSA and the found parameters from the experiments are comparable for low ΔT, given the restrictions of the LSA and experimental possibilities in the parabolic flight. However, the first experimental transitions leads apparently to a columnar pattern. The onset of helical pattern can be found in $1g$, but none of these patterns can be confirmed with high confidence. It is possible that the columns are slanted with a very low pitch which cannot be resolved in the images. In this case it might be possible to acquire better results simply by changing the position of the LLS. The curve of $\gamma_{e,crit}$ with a plateau at $\gamma_e \approx 0.015$ for $V_E < 420$ is not found in the experimental data. The critical voltage for $\gamma_e > 0.01$ ($\Delta T > 10K$) becomes about $V_E = 1100 \cdots 1300$ ($V_P = 6kV \dots 7kV$).

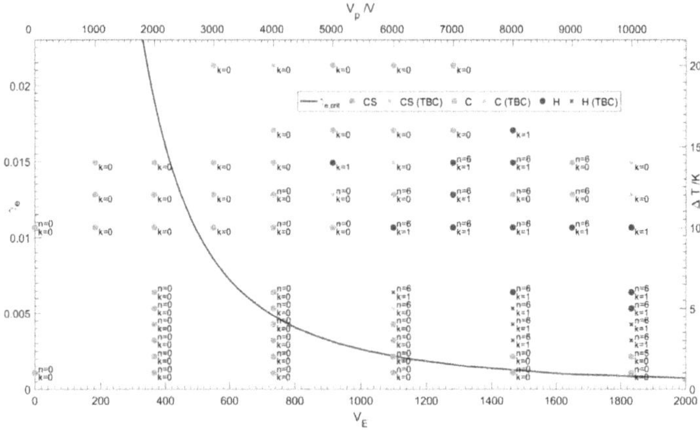

Figure 7.25.: Comparison of the LSA and experiments in μg . The blue line shows the critical parameters calculated by the LSA. CS - conductive state, C - columnar pattern, H - helical pattern, TBC - to be confirmed.

The experimental and theoretical values for the μg phase also only fit for low ΔT. The higher the temperature, the bigger the difference. In contrast to the $1g$ data, the pattern after the first transition is a helical pattern, or the transition area showing a columnar pattern is smaller. Less external force is required to establish a helical pattern. The critical voltage for $\gamma > 0.01$ ($\Delta T > 10K$) is decreased during μg and becomes about $V_E = 900 \ldots 1100$ ($V_p = 5kV \ldots 6kV$).

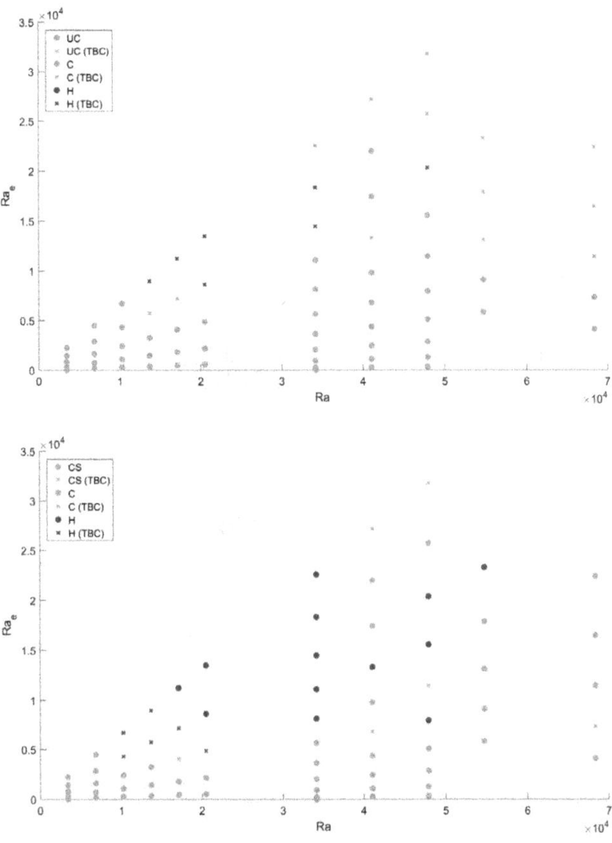

Figure 7.26.: Comparison of the LSA and experiments in $1g$ (top) and μg (bottom) given as $\mathcal{R}a$ and $\mathcal{R}a_e$. The blue line shows the critical parameters calculated by the LSA. UC - unicellular pattern, CS - conductive state, C - columnar pattern, H - helical pattern, TBC - to be confirmed.

7.4. Heat transfer and Nusselt number

The Nusselt number $\mathcal{N}u = \frac{\text{Total heat transfer}}{\text{Conductive heat transfer}}$ is an indicator for the efficiency of convective heat transfer. If $\mathcal{N}u = 1$ then there is only conductive heat transfer and the higher it is the higher is the convective heat transfer. It is calculated as given by equation 3.19. The temperature measurements should be treated as qualitative direction in which the temperature and the heat transfer develops. The thermal relaxation time of the fluid is about 5 minutes (table A.6), which is way longer than the 22s of μg time. When the temperature difference is high enough ($\Delta T > 10K$) it is possible to obtain repeatable results for the measurements and $\mathcal{N}u$ calculations. However, for a quantitative evaluation it is required to have a longer time in μg , such as on the ISS or a TEXUS mission. The actual measured fluctuations of the temperature within the μg time are approx. $0.3K$ for the $100mm$ cell, which is very close to the error range of $0.2K$. For lower temperature gradients ($\Delta T < 10K$), the measured temperature changes are lower than the measurement error and are hence not presented here.

The temperature data is filtered using a Savitzky-Golay filter (Savitzky *et al.* 1964). This approach performs a convolution of the signal to determine a polynomial. This improves the signal-to-noise ratio without eliminating local minima and maxima and without introducing a delay like a moving average would do. The temperature gradient over the gap is calculated by $T_{H,i} - T_{C,i}$. The heating and cooling loop are in a counterflow configuration. The volume flow rate is compensated for the temperature dependency of the density.

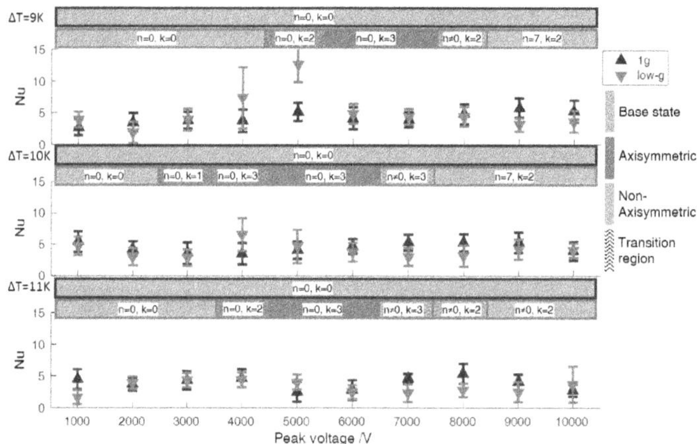

Figure 7.27.: Nusselt numbers for different ΔT and V_p in the $30mm$ cell. The base state is a unicellular flow in $1g$ and a conductive state in μg . Since the flow pattern cannot be classified with high confidence in some cases, they are simply categorized as axisymmetric or non-axisymmetric. $\mathcal{N}u$ given in the $1g$ phase is a mean over the last $10s$ before the $1.8g$ phase starts. $\mathcal{N}u$ given in the μg phase is a mean over the last $4s$ before the μg phase ends.

The Nusselt numbers are presented as a scatter plot including the standard deviations error bars. Above the plot are colored bars to indicate the observes flow pattern including the mode numbers. In some cases it is not possible to determine this and I can only indicate whether it is $= 0$ or $\neq 0$. The base state is, again, depending on g_z. In a $1g$ environment it is a unicellular flow and in μg it is a conductive state.

The results for the $30mm$ cell are given in Fig. 7.27. The flow pattern does not change in $1g$ based on the used visualization methods, but $\mathcal{N}u$ changes when the voltage is changed. Since the data was taken during the $1g$ phase of the parabolic flights, there might be an influence by the changing g_z levels, although the $1g$ phase is quite long compared to the other gravity phases. The changes seen during the μg phase, when also the flow pattern changes, are more noticeable. At $\Delta T = 9K$ are two voltages, $V_p = 4kV$ and $V_p = 5kV$, where the absolute $\mathcal{N}u$ and the error suddenly increases. The flow pattern starts to change from the conductive state to an axisymmetric pattern at these parameters. It is possible that the flow is still in a transition state, which has a higher heat transfer than a more stable state e.g. at $v_p = 6kV$. There is exactly the same phenomenon at $\Delta T = 10K$. Again, the flow pattern starts to change at these parameters and the time in μg is too short to get reliable results. The series for $\Delta T = 11K$ shows only an outlier at $V_p = 10kV$, where the error of the μg data suddenly increases. Overall it seems that $\mathcal{N}u$ decreases at higher voltages ($V_p > 7kV$) in μg compared to $1g$. It is possible that the flow pattern created at these parameters inhibit the heat transfer. At lower voltages there is mostly no difference between μg and $1g$ phase, which means that the difference in heat transfer is marginal.

Similar to the figure for the $30mm$ cell there is also one for the $100mm$ cell (Fig. 7.28). At $\Delta T = 10K$, $\mathcal{N}u$ is rather constant until $V_p = 6kV$. After this point $\mathcal{N}u$ increases slightly. This is also the point at which the flow pattern changes. At $V_p = 5kV$ is an increase of the error in the μg phase visible. The flow could still be in transition and cannot stabilize in the limited μg time. At $\Delta T = 12K$ are more parameters at which $\mathcal{N}u$ has a larger error. At $V_p = 4kV$, $V_p = 5kV$ and $V_p = 7kV$ it can be explained by the changing flow pattern. However, the error at $V_p = 1kV$ cannot be explained by looking at the experimental data. The $\mathcal{N}u$ numbers at $\Delta T = 14K$ are comparable to the ones at $\Delta T = 10K$, but slightly higher. Again, the error increases at the transition of the flow pattern at $V_p = 7kV$. Also, $\mathcal{N}u$ seems to increase slightly after the flow pattern changed.

Based on the results obtained from the PFC, I would assume that the effect on the convective heat transfer in the $30mm$ cell is not very beneficial. The data does not show a clear increase of $\mathcal{N}u$ with increasing voltage independent of g_z. But one should remember that the measured differences in the temperature is very low ($< 0.1K$) for this cell. The calculated $\mathcal{N}u$ should hence be treated with care. Performing experiments with longer μg phase would very likely lead to different results. The data obtained from the $100mm$ cell is quantitatively comparable to the data by given in Futterer, Dahley, *et al.* 2016. $\mathcal{N}u$ seems to be constant or even decrease slightly until a critical voltage is reached. After that $\mathcal{N}u$ increases. With this results it is possible to find a connection between $\mathcal{N}u$ and the flow pattern. In both, this work and the work done by Futterer et al., $\mathcal{N}u$ starts to increase between $V_p = 6kV$ and $V_p = 7kV$, depending on ΔT and

g_z. This is also the parameter when the flow changes from a base state to a different pattern. But again, more precise determinations of $\mathcal{N}u$ in the $100mm$ cell are required, which can only be done in long term μg experiments.

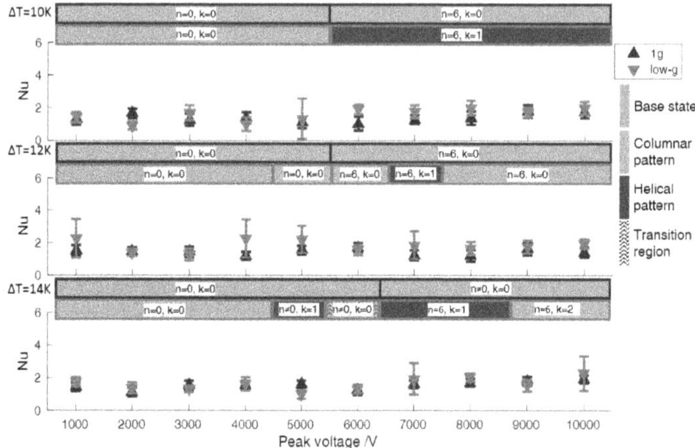

Figure 7.28.: Nusselt numbers for different ΔT and V_p in the $100mm$ cell. The base state is a unicellular flow in $1g$ and a conductive state in μg . $\mathcal{N}u$ given in the $1g$ phase is a mean over the last $10s$ before the $1.8g$ phase starts. $\mathcal{N}u$ given in the μg phase is a mean over the last $4s$ before the μg phase ends.

The temperature sensors glued into the outer cylinder of the type T cells measure the local temperatures along a defined axial and azimuthal line. This information could help to identify the flow pattern, e.g. by making it possible to see if a column is slanted or not, and could increase the accuracy of the temperature measurements because more than one sensor is used. This idea worked on a global level, but did unfortunately not work on a local level. The temperatures only change by $< 0.1K$ during the μg phase. This is too low to give reliable results with the existing measurement system. Due to the high electric field it is not possible to move the sensors directly into flow inside the gap. They have to be inside the outer cylinder to prevent influences of the electric field on the measurements. This reduces the sensitivity of the sensors towards the temperature of the fluid in the gap.

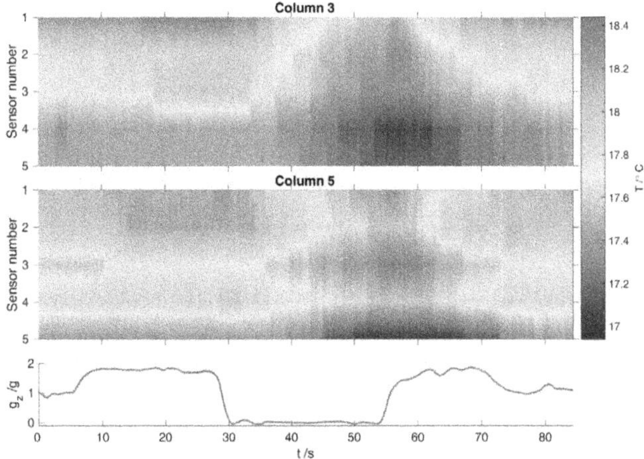

Figure 7.29.: Space-time plots of the temperatures at column 3 and 5. The experimental parameters are $\Delta T = 16K$, $V_p = 5kV$ and $H = 30mm$. The sensors are aligned to the axis of the cell, where sensor 1 is at the top and sensor 5 at the bottom. Both plots should show the qualitatively same, but they differ in the μg phase. The positions of the sensors are given by the matrix in Fig. 5.4.

An example for such a temperature field, which is a space-time plot of the temperature, is given in Fig. 7.29. The measurements have been performed with the $30mm$ cell at $\Delta T = 16K$ and $V_P = 5kV$. Based on the evaluations of the other visualization methods it should be an unicellular pattern in the $1g$ phase and likely a non-axisymmetric pattern in the μg phase. The plots show the temperature along two different lines along axial direction (Fig. 5.4). Sensor 1 is the topmost and sensor 5 is at the bottom. For the $1g$ and $1.8g$ phase the plot shows a gradient from warm (top) to cold (bottom). This is a result of the unicellular flow. It can be seen clearly at column 3, but not so clearly at column 5. After the start of the μg phase at $t = 30s$ the temperature decreases a bit. The temperature at column 3 decreases homogeneously, which would be the case of a conductive state. However, column 5 shows a different pattern. There are alternating warmer and colder regions. Compared to the results from the PIV images, this would be caused by toroidal vortices.

The temperature field shows some contradictions. At one point in the gap it suggests, that an unicellular pattern exist in $1g$ and a conductive state in μg . At a different azimuthal position is an unicellular pattern in $1g$ and a probably axisymmetric pattern in the μg phase. The spatial distribution of this approach is too low to give reliable results. Also the low thermal diffusion time of the silicone oil reduces the usefulness of this method in parabolic flights.

7.5. Effect of the frequency of the electric field

One of the basic assumptions for the work in this thesis is that the Coulomb force can be neglected and only the dielectrophoretic force acts on the fluid. This means that the frequency of the electric field needs to be way higher than the inverse of the electric relaxation time of the fluid. This relaxation time is given by

$$\tau_e = \frac{\varepsilon_0 \varepsilon_r}{\sigma} \tag{7.1}$$

with σ the electric conductivity and $\varepsilon_0\varepsilon_r$ the absolute permittivity of the fluid (Takashima 1980). For AK5, $\tau_e = 23.9s$. This means one needs to satisfy $f >> 0.0418Hz$. However, there is no rule or estimation when this is satisfied. For AK0.65 $\tau_e = 19.3s$ ($f >> 0.0518Hz$) and Novec 7200 $\tau_e \approx 2.6s$ ($f >> 0.3846Hz$). The critical frequencies for AK5 and AK0.65 are quite similar, but experiments have shown that different excitation frequencies are required.

7.5.1. Shadowgraph method

In this experimental run I kept the temperature gradient and peak voltage constant ($\Delta T = 10K$, $V_p = 10kV$), but increased the frequency of the a.c. field step-wise. It was visualized using the Shadowgraph method. This gives better results than PIV, because the particles could behave differently than the fluid when subjected to different frequencies. To eliminated this factor I did not chose to use PIV as measurement method. The results are given as space-time plot at a constant radius near the outer cylinder where the resulting structures are visible. There is an alternating pattern over time visible in all of the plots. In experiments I found that this effect is caused by the cameras themselves and is probably an artifact caused by the rolling shutter (see Appendix A.1). The pattern exists also in the results of the PFC experiments, but is less apparent because the timescale is lower.

At $f = 0.1Hz$ (Fig. 7.30) there is a clear shadow visible whenever the field changes polarity. This is caused by the fluid which is moved in bulk when the polarity changes. This effect is caused by the Coulomb force, which seems to be non-negligible although the frequency is already higher than theoretically needed. I assume that the effect of the Coulomb force is negligible small when this effect is not visible anymore.

At $20Hz$ (Fig. 7.31) the pattern changes to a columnar pattern. However, it changes over time and the position of the columns reorganize. It seems to be stable after about 7 minutes. When looking at the full images as a video sequence it seems that the mode number increases from $n = 5$ to $n = 6$. It seems that $20Hz$ is sufficient, but the growth rates of the instabilities are very low. I assume that the DEP force is not yet strong enough compared to the other forces.

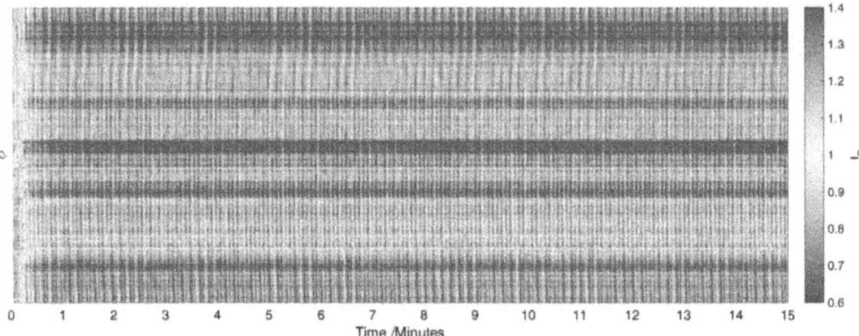

Figure 7.30.: Space-time plot of a Shadowgraph image at a constant radius with $\Delta T = 10K$, $V_p = 10kV$ and $f = 0.1Hz$. The visualization shows a pattern which pulsates whenever the voltage changes polarity.

The next step was at $f = 40Hz$ (Fig. 7.32). Starting from this frequency the resulting Shadowgraph image always showed a columnar flow pattern with $n = 6$. The position of the columns still changes very slightly over time. At frequencies of $60Hz$ and higher there is no further change of the position of the columns. Based on this results I determined the minimal required frequency for AK5 to be $60Hz$.

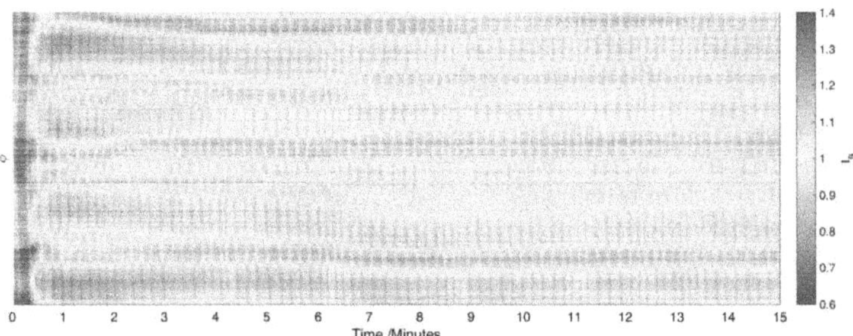

Figure 7.31.: Space-time plot of a Shadowgraph image at a constant radius with $\Delta T = 10K$, $V_p = 10kV$ and $f = 20Hz$. A columnar pattern establishes and rearranges itself over about 7 minutes.

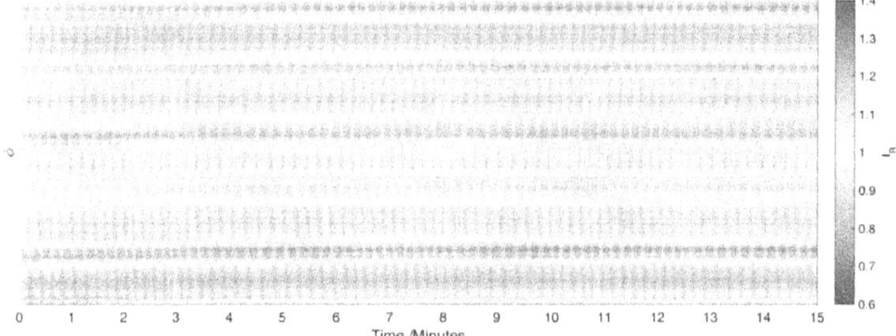

Figure 7.32.: Space-time plot of a Shadowgraph image at a constant radius with $\Delta T = 10K$, $V_p = 10kV$ and $f = 40Hz$. The columns arrange very slightly, but the mode does not change.

7.5.2. PIV particles

A similar experimental run was performed to determine the behavior of the PIV particles inside the fluids AK5, AK0.65 and Novec 7200. In most cases is the permittivity of the particles higher than the permittivity of the fluids. This means that the particles are more sensitive to the electric field than the fluid. For this experiments, the temperature gradient is fixed at $\Delta T = 10K$ with a voltage amplitude of $V_p = 10kV$ and a variable frequency in the $H = 100mm$ cell. The frequency must be high enough that the effect of the Coulomb force can be neglected for the fluid and the particles. This and the slip was calculated by Seelig *et al.* 2019. The slip describes whether the particles follow the flow or not based on the friction between the fluid and the particles. He came to the conclusion that the particles follow the flow very well for the conditions of the experiment. In the experiments with AK5 the particles seem to follow the flow for $f = 200Hz$, so no further research was done to this since it already fulfills the requirements. The results are presented as space-time plot of the untreated PIV images which show the particles. Two positions along the height are chosen. One near the inner cylinder and one near the outer.

For AK0.65 the electric field has a stronger effect on the particles than on the fluid. As soon as the electric field is activated the particles move towards the inner cylinder. This happens independent from the frequency and can be seen very well at lower frequencies like $0.1Hz$ (Fig. 7.33). The voltage is activated at $t = 2s$. The particles move upwards near the heated inner cylinder and downwards near the cooled outer cylinder before that point of time. After that a movement in radial direction is visible. The particles move from one side of the cylinder to the other side whenever the voltage changes polarity.

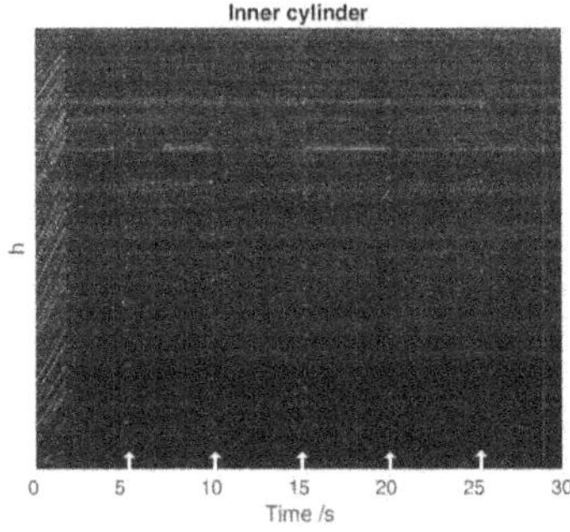

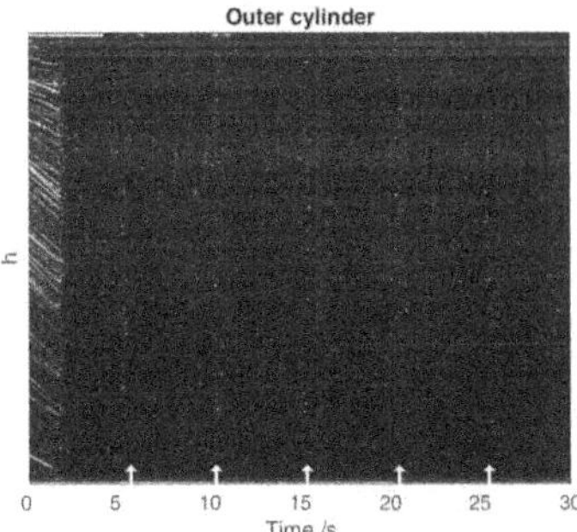

Figure 7.33.: Space-time plot of an untreated PIV image at a constant radius with $\Delta T = 10K$, $V_p = 10kV$ and $f = 0.1Hz$ activated at $t = 2s$. An oscillating pattern is visible. The particles move from one side to the other whenever the voltage changes polarity.

An oscillatory pattern is still visible when a frequency of $f = 200Hz$ is used. However, this is limited to the inner cylinder. The particles at the outer cylinder show an effect of the electric field, but also seem to follow the flow. The electric field is non-homogeneous due to the geometry and stronger near the inner cylinder (cf. Fig. 1.4). However near the inner cylinder, the particles are still influenced strongly by the field. This effect can still be found at $f = 400Hz$ (7.35).

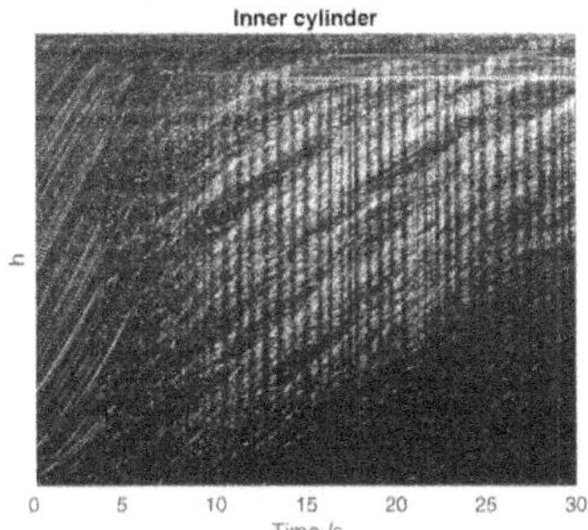

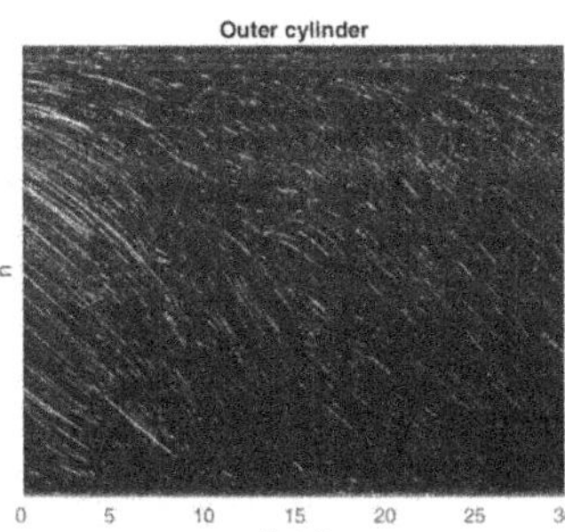

Figure 7.34.: Space-time plot of an untreated PIV image at a constant radius with $\Delta T = 10K$, $V_p = 10kV$ and $f = 200Hz$activated at $t = 7s$. An oscillating pattern is visible near the inner cylinder, but not near the outer cylinder.

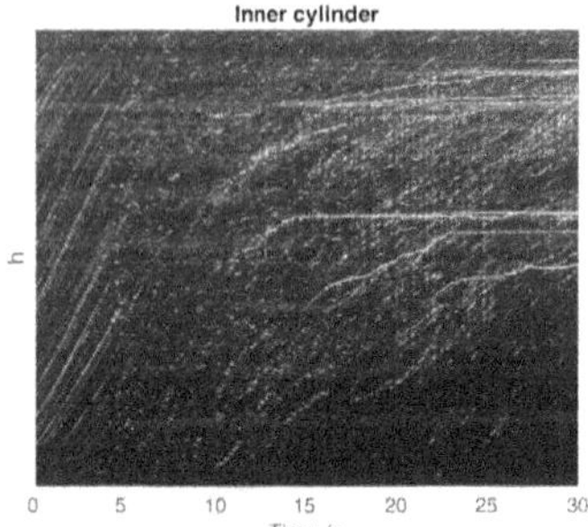

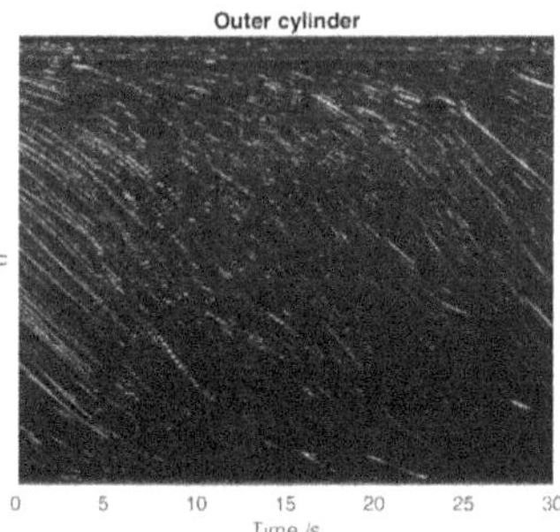

Figure 7.35.: Space-time plot of an untreated PIV image at a constant radius with $\Delta T = 10K$, $V_p = 10kV$ and $f = 400Hz$ activated at $t = 5s$. An oscillating pattern is visible near the inner cylinder, but not near the outer cylinder.

The result of the experiment series with AK0.65 was that we decided to use $f = 500Hz$ for the fluid and particle combination. This is still at the low-end, but due to technical limitations of the high-voltage amplifier it was not possible to increase the frequency further without deforming the sine wave. The experiments with Novec 7200 were performed as pre-experiments for the plate cavity project, but led to a similar result. Both of these fluids, AK0.65 and Novec 7200, have a lower viscosity than AK5. There is less interaction between the fluid and the particles, which increases the particle slip. The particles are more likely to follow the electric field than the flow of the fluid.

7.6. Effects on other dielectric fluids

The results presented in this thesis were acquired mainly from experiments with AK5. But I also looked at other fluids. Several experiments were performed with AK0.65, which is also a silicone oil, but has a lower viscosity than AK5, and Novec 7200, which is an alcohol-based fluid engineered for application as heat transfer fluid. Their properties can be found in Table A.6.

7.6.1. Novec 7200

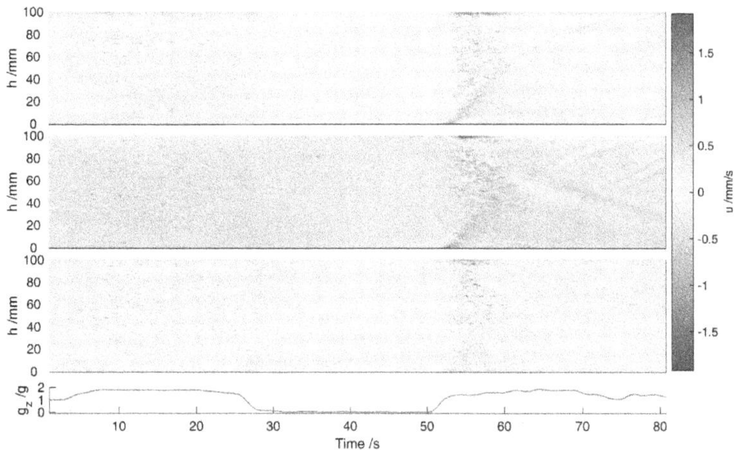

Figure 7.36.: Space-time plot of the radial velocity of Novec 7200 at $\Delta T = 1K, V_p = 0kV$ at different radii (top: near inner cylinder, middle: middle of the gap, bottom: near outer cylinder). Without external forces there is no movement of the particles nor the flow during the μg phase.

Novec 7200 was used during a PFC in 2018 in the PIV cells. The used particles were made from polyamide P84-NT1. Since this fluid has a high ε_r and e, lower temperature gradients and voltages are required. The voltage is active over the whole experimental run unless stated otherwise. The results are presented as space-time plot of the radial velocity at different locations. One plot is done near the inner cylinder ($r = 6mm$), one at the middle of the gap ($r = 7.5mm$) and one near the outer cylinder ($r = 9mm$). As shown in chapter 7.5 the influence of the electric field on the particles depends on the position in the gap, which can make a difference for low viscous fluids.

As reference measurement a temperature gradient $\Delta T = 1K$ was applied without electric field (Fig. 7.36). This shows the expected unicellular pattern in $1g$ and no particle movement in the μg phase. The transition in the second $1.8g$ phase has a different flow pattern than AK5. Some vortices seem to establish near the inner cylinder and move upwards. This structure disappears after reaching the $1g$ phase again. The flow pattern needs more time to reach a stable state because Novec 7200 has a lower viscosity.

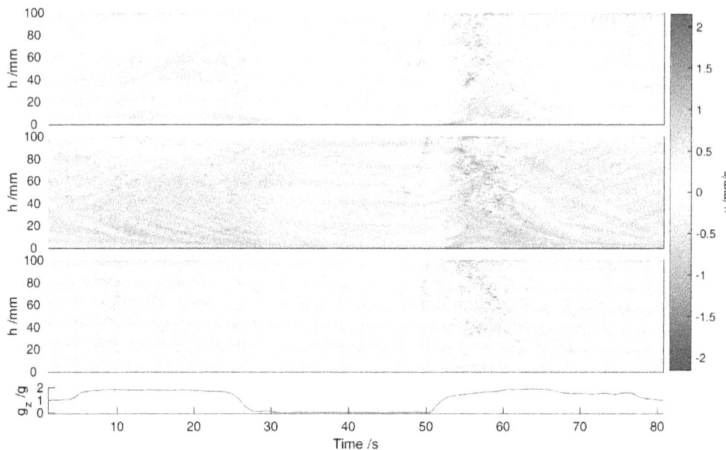

Figure 7.37.: Space-time plot of the radial velocity of Novec 7200 at $\Delta T = 1K, V_p = 0.1kV$ at different radii (top: near inner cylinder, middle: middle of the gap, bottom: near outer cylinder). A certain axisymmetric structure can be seen in all phases near the inner cylinder and in the middle of the gap. At the outer cylinder, the flow pattern is not axisymmetric.

At $\Delta T = 1K$ with a peak voltage of just $V_p = 0.1kV$ it is possible to see a change in the flow pattern (Fig. 7.37). Near the outer cylinder seems to be no effect of the voltage. This could, again, be caused by the inhomogeneous strength of the electric field. It seems that there is an axisymmetric pattern at the inner cylinder in the $1g$ phase. However, it cannot be seen at the other locations. In the $1.8g$ phase it changes to an unicellular pattern. In the μg phase it changes to an axisymmetric pattern near the inner cylinder again. The transition to $1.8g$ seems to be turbulent and again axisymmetric in the $1g$ phase. The pattern could be a complex superposition of an axisymmetric pattern at the inner part of the cell and a non-axisymmetric pattern at the outer part. With the applied measurement technique it is not possible to determine the pattern precisely. The pattern which can be seen at the top of the images starting at about $t = 45s$ is caused by a bubble in the fluid.

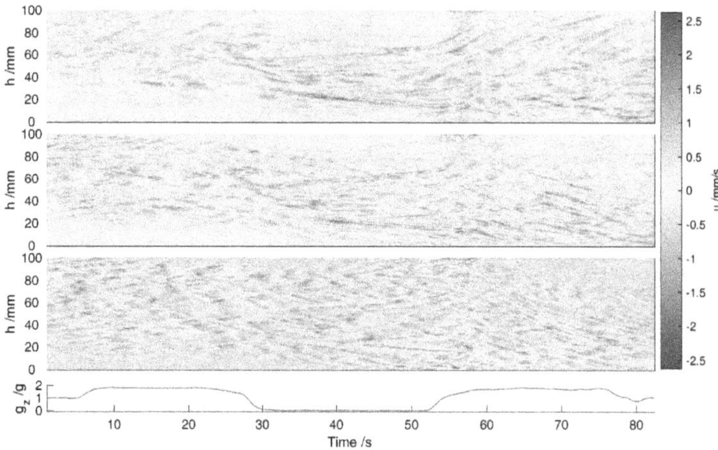

Figure 7.38.: Space-time plot of the radial velocity of Novec 7200 at $\Delta T = 1K, V_p = 0.2kV$ at different radii (top: near inner cylinder, middle: middle of the gap, bottom: near outer cylinder). The flow pattern seems to be turbulent.

Increasing the voltage further (e.g. $V_p = 0.2kV$, Fig. 7.38) changes the flow pattern to be turbulent. The higher the voltage, the more turbulent it becomes.

In an additional measurement series the temperature gradient was fixed at $\Delta T = 4K$ and the voltage as well as the time at which the voltage was activated changed. For the presented data $V_p = 0.5kV$. When the voltage is active over the whole experimental run (Fig. 7.39) the pattern is similar to the one at $\Delta T = 1K, V_p = 0.1kV$ (Fig. 7.37). During the first $1g$ and $1.8g$ phases an axisymmetric pattern is visible near the inner cylinder and the middle of the gap. During the μg phase this changes to a different pattern. After a turbulent transition phase it becomes axisymmetric again.

When the same parameters are applied, but the voltage is activated in the middle of the μg phase ($t = 40s$) and active for the rest of the experimental run (Fig. 7.40), the pattern changes a bit. The pattern near the inner cylinder and in the middle of the gap disappears during the first $1g$ and $1.8g$ phases. Since there is no voltage, there is no radial force acting on the flow and the particles. During the first half of the μg phase there is no visible movement. As soon as the voltage is activated at $t = 40s$ a pattern starts to emerge. While it appears to be axisymmetric at the upper and lower boundaries it is not possible to determine the pattern with a high confidence. During the second $1.8g$ and $1g$ phases, when the voltage is still active, the pattern looks exactly like in the previous experimental run (Fig. 7.39).

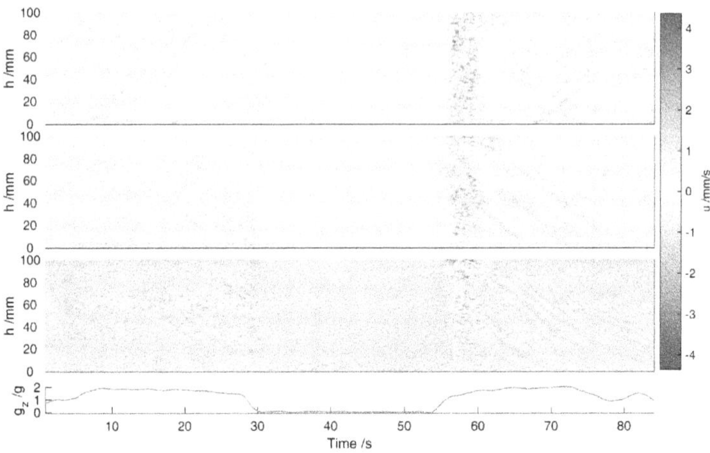

Figure 7.39.: Space-time plot of the radial velocity of Novec 7200 at $\Delta T = 4K, V_p = 0.5kV$ at different radii (top: near inner cylinder, middle: middle of the gap, bottom: near outer cylinder). An axisymmetric pattern can be seen in the $1g$ and $1.8g$ phases while the pattern in the μg phase is non-specific.

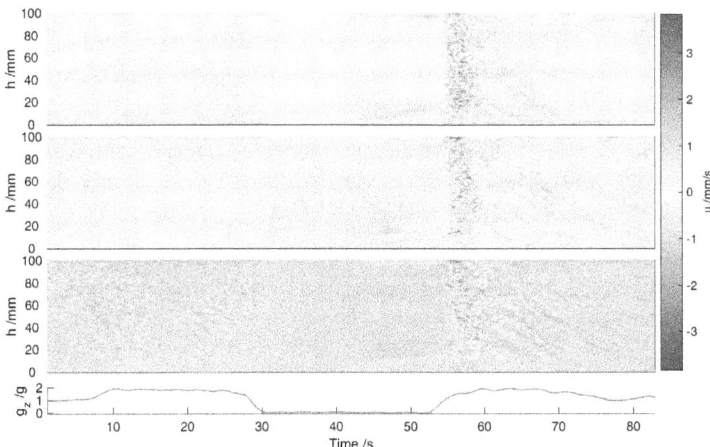

Figure 7.40.: Space-time plot of the radial velocity of Novec 7200 at $\Delta T = 4K, V_p = 0.5kV$ at different radii (top: near inner cylinder, middle: middle of the gap, bottom: near outer cylinder). The voltage is activated at $t = 40s$ and stays active until the end of the graph. There seems to be no specific pattern in the first $1g$ and 1.8 phases, and the μg phase. But the second $1g$ and $1.8g$ phases, when the voltage is active, suggest an axisymmetric pattern.

7.6.2. AK0.65

The experiments for AK0.65 were performed with the Synthetic Schlieren technique in the $30mm$ cell. A lot of different ΔT and V_p combinations were examined, but only a small set of this is shown. The presented data was acquired during my second PFC in 2016 and the measurement setup has been improved since then. The acquired flow patterns differ a lot from the patterns seen with AK5. Similar to Novec 7200 there are different effects near the inner and outer cylinders. The results are given as space-time plot of the displacement given by the Synthetic Schlieren images at different radii. One is near the inner cylinder, on in the middle of the gap, and one near the outer cylinder.

In the presented data the temperature difference is always $\Delta T = 7K$ and the voltage is changed. The lowest applied voltage is $V_p = 2kV$ (Fig. 7.41). These are critical parameters as predicted by a linear stability analysis (Yoshikawa *et al.* 2013) and should produce perturbations in the flow. The flow pattern at the middle of the gap and near the outer cylinder show a homogeneous change when entering the μg phase, which indicates that there is no specific flow pattern, but probably a conductive state. But near the inner cylinder it is possible to see the onset of a inhomogeneous structure.

When the voltage is increased to $V_p = 4kV$ (Fig. 7.42) one can see that this structure develops also at the other locations inside the gap. At $V_p = 6kV$ (Fig. 7.43) it develops at the outer cylinder, but the flow pattern near the inner cylinder seems to develop to a different pattern. The pattern near the outer cylinder at the end of the μg phase can be compared to the columnar pattern seen with AK5, while the pattern near the inner cylinder seems to show an oscillatory effect. This effect is unfortunately not visible well on these space-time plots, but only in the videos. It could also be an artifact caused by a strong blurring of the pattern. Due to the strong blur the evaluation algorithm becomes less reliable and can cause errors.

As the voltage is increased even further to $V_p = 8kV$ (7.44) or $V_p = 10kV$ (7.45) the pattern starts to disappear. The pattern near the inner cylinder seems to become more turbulent. The pattern at the middle of the gap and near the outer cylinder seems to become turbulent as well. However, based on just these images taken during a parabolic flight it cannot be determined precisely.

The observed pattern changes slightly in nearly all images when the gravity changes from $1g$ to $1.8g$. This is expected since the increased gravity increases the thermal convection. When entering the μg phase the change of the pattern is more obvious, since the flow state changes completely. Due to the low viscosity of AK0.65 it requires a longer time to reach a stable state than AK5 (cf. Table A.6). While the measurements during parabolic flights allow to confirm that there is an effect of the DEP force on this fluid, it is not possible to see a stationary state and compare it to the theory.

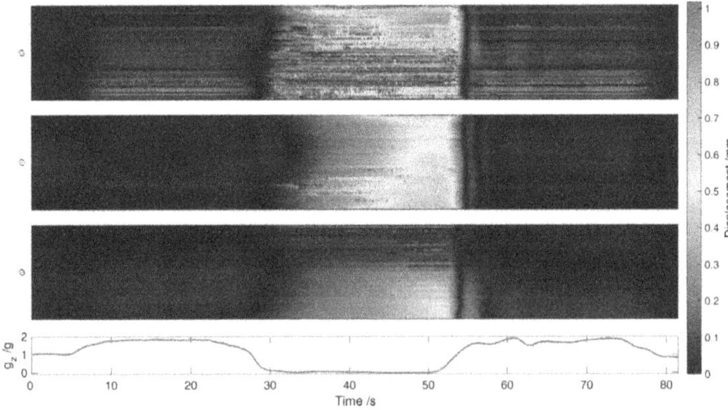

Figure 7.41.: Space-time plot of the displacement of the pattern in the Schlieren images at $\Delta T = 7K, V_p = 2kV$ in the $30mm$ cell at different radii (top: near inner cylinder, middle: middle of the gap, bottom: near outer cylinder).

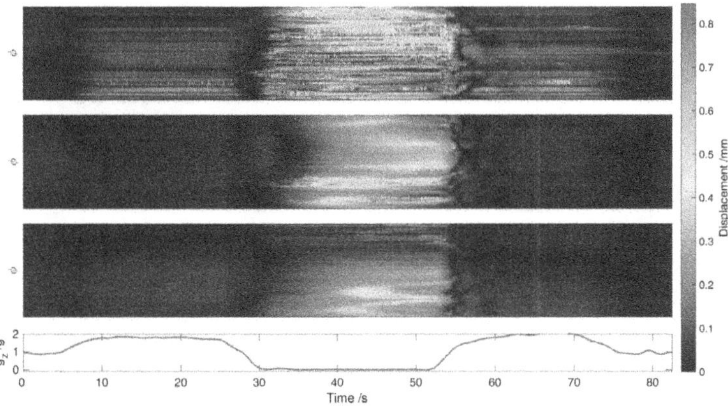

Figure 7.42.: Space-time plot of the displacement of the pattern in the Schlieren images at $\Delta T = 7K, V_p = 4kV$ in the $30mm$ cell at different radii (top: near inner cylinder, middle: middle of the gap, bottom: near outer cylinder).

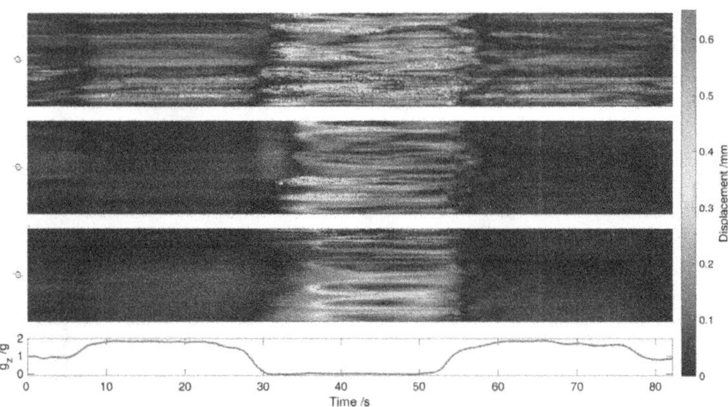

Figure 7.43.: Space-time plot of the displacement of the pattern in the Schlieren images at $\Delta T = 7K, V_p = 6kV$ in the $30mm$ cell at different radii (top: near inner cylinder, middle: middle of the gap, bottom: near outer cylinder).

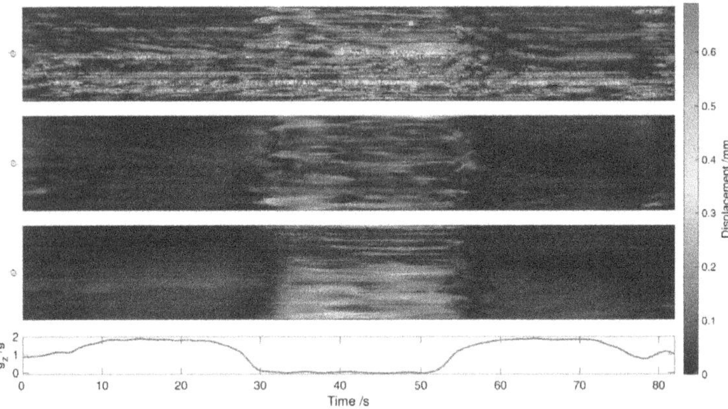

Figure 7.44.: Space-time plot of the displacement of the pattern in the Schlieren images at $\Delta T = 7K, V_p = 8kV$ in the $30mm$ cell at different radii (top: near inner cylinder, middle: middle of the gap, bottom: near outer cylinder).

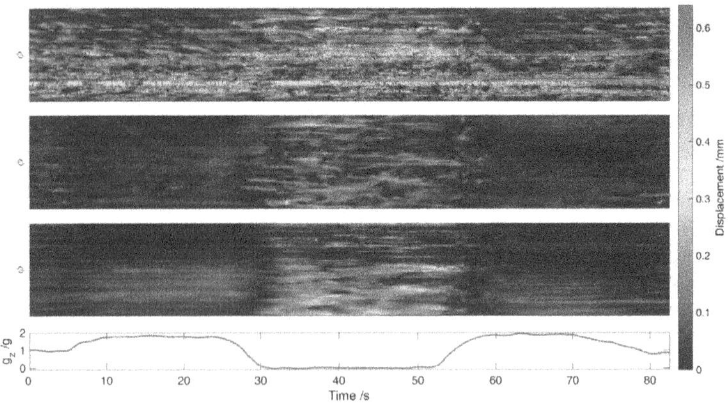

Figure 7.45.: Space-time plot of the displacement of the pattern in the Schlieren images at $\Delta T = 7K, V_p = 10kV$ in the $30mm$ cell at different radii (top: near inner cylinder, middle: middle of the gap, bottom: near outer cylinder).

To make a qualitative comparison between the development of the columnar flow pattern in AK5 and AK0.65, different experimental parameters are compared. Figure 7.46 shows the development of a columnar pattern at $\Delta T \approx 10K$ and $V_p = 10kV$ using the 30mm cell and AK5. The first $1g$ and $1.8g$ phases, where the voltage is not active, show a homogeneous pattern. Some artifacts can be seen caused by the measurement setup and used pattern at this point in time. The voltage is activated $10s$ into the μg phase and about $5s$ later a columnar pattern establishes. This pattern is still visible during the second $1g$ and $1.8g$ phases because the voltage is still active. Once the basic shape of this pattern has developed only the position adjusts a bit, but the shape stays the same.

A qualitatively comparable pattern can be seen in AK0.65 in the 30mm cell at $\Delta T \approx 10K$ and $V_p = 7kV$ (Fig. 7.47). The voltage is also activated $10s$ into the μg phase and the first $1g$ and 1.8 phases also show a homogeneous pattern similar to AK5. After the voltage is activated a columnar pattern similar to AK5 appears near the outer cylinder. Near the inner cylinder and at the middle of the gap it seems at first that also a columnar pattern starts to develop. But instead of becoming stable it seems to be more prone to oscillatory effects. This could also be caused by the low viscosity of the fluid and the strong electric field at the inner cylinder.

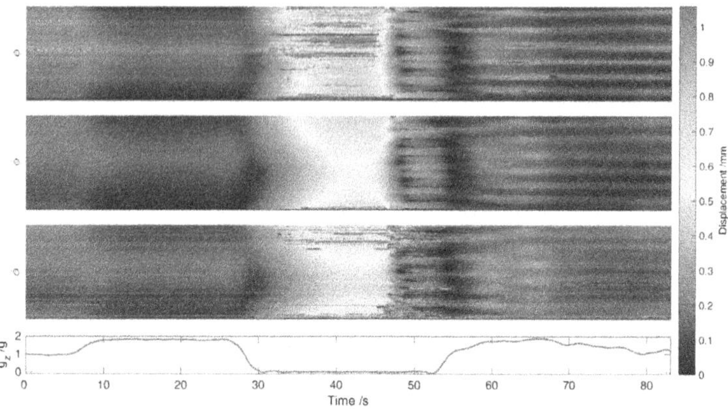

Figure 7.46.: Space-time plot of the displacement of the pattern in the Schlieren ¡images at $\Delta T = 10K, V_p = 10kV$ in the $30mm$ cell filled with AK5 at different radii (top: near inner cylinder, middle: middle of the gap, bottom: near outer cylinder). The voltage is activated $10s$ into the μg phase ($t = 40s$).

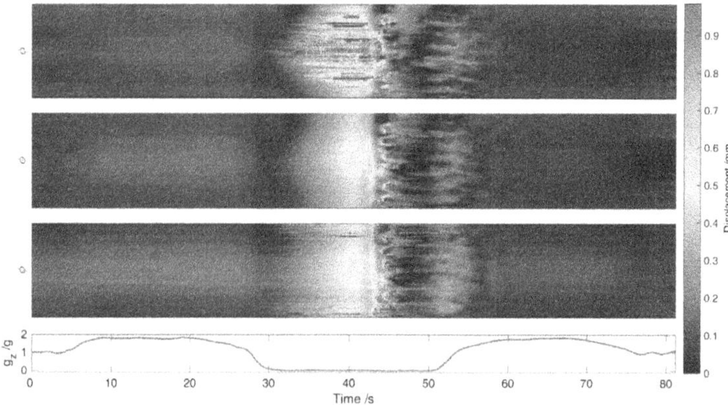

Figure 7.47.: Space-time plot of the displacement of the pattern in the Schlieren images at $\Delta T = 10K, V_p = 7kV$ in the $30mm$ cell filled with AK0.65 at different radii (top: near inner cylinder, middle: middle of the gap, bottom: near outer cylinder). The voltage is activated $10s$ into the μg phase ($t = 40s$).

8. Discussion

During the research period on this topic, I performed experiments with a small cell with $H = 30mm$ and one larger cell with $H = 100mm$. The main results presented in this thesis have been acquired with silicon oil AK5 as working fluid, but other fluids have been tested. Over 400 experiments have been performed under μg conditions in parabolic flight campaigns with different parameters, fluids and measurement techniques. The combination of several experimental runs are used to determine the structure of the flow and the heat transfer.

The flow inside the $30mm$ cell with $\Gamma = 6$ is stabilized by the boundary effects. In the $1g$ phases of the parabolic flight no perturbations are visible (Fig. 7.11). Even at high voltages the flow always an unicellular pattern. During the μg phases, several different pattern are found (Fig. 7.12). The first instability is a mostly toroidal pattern, which is not always fully developed. A second instability, which appears at higher voltage, also introduces columnar-like structures. These structures always show a toroidal pattern at the boundaries. Due to the large influence of the boundary effects and limitations of the experimental setup and PFC, I am not able to qualify this pattern with high confidence. The patterns are generalized as axisymmetric and non-axisymmetric pattern. Since the LSA assumes a cell of infinite length, where boundary has no effect, it is not feasible to make a comparison to the $30mm$ cell, which is dominated by boundary effects. Contrary to the predictions of the LSA I was not able to find any perturbations in $1g$. The boundary effects stabilize the flow, so that the influence of the electric field is too weak. However, in the μg phase the critical parameters given by the LSA can be found in the experiments as well. The critical parameters fit very well for $\Delta T < 3K$. The higher ΔT, the higher the divergence between the LSA and the experiments. Since the LSA applies Boussinesq approximations it is not surprising to see a divergence at high ΔT. The LSA predicts a helical pattern for the flow when the critical parameters are reached. This pattern cannot be found in the $30mm$ cell. The first observed stability is toroidal. A second instability is a columnar pattern located in roughly the middle third of the gap height and toroidal vortices at the boundaries. However, it is not possible to determine whether the column is slanted or not.

The boundary effects have a lower influence on the flow in the $100mm$ cell. It is possible to differentiate between a base state, columnar, and helical pattern, with the exception of some parameters, where the flow is still in transition. The difference of the columnar and helical pattern is given by the vectorfield of the PIV images. The radial component of the vectors points in the same direction for columnar vortices. For helical vortices there is a gradient in the radial component, which also changes the sign. Depending on the pitch of the helical columns, i.e. the angle at which the column is inclined, and the position of the LLS it could be possible to miss the change in sign.

Toroidal vortices are only found at the boundaries, but never spanning the whole gap. When the voltage is activated at the start or in the middle of the μg phase, toroidal vortices can be found near the middle of the gap (Fig. 7.20 and 7.23). Under this conditions there can also be more toroidal vortices near the boundaries than when the voltage is active continuously. In $1g$, the LSA predicts a first change of the flow pattern from the unicellular flow to a columnar flow at $V_{E,crit1} = 425.8$ ($V_{p,crit} = 2.3kV$). A second transition to a helical pattern occurs at $V_{E,crit2} = 3796$ ($V_{p,crit2} = 14.2kV$). The second critical voltage is outside of the possible parameters of the experiment. The experiments also show a columnar pattern after a first transition. But this occurs usually at higher voltages than expected from the LSA, even at low ΔT. The data is acquired during the 1 minute long $1g$ phases of the PFC. This may not be enough time for the flow to stabilize. Laboratory experiments concerning this topic have been performed by Torsten Seelig, who works in our TEHD research group, with the same experimental cells and similar parameters. The experiments were only performed in $1g$, but the experimental runs last 15 minutes instead of 1 minute. The critical voltages are lower in his results and they fit very well for $\Delta T < 3K$ (Seelig *et al.* 2019). In contrast to the PFC data, the $1g$ data also does not suggest a helical pattern at any point. The continuously changing g_z may influence the flow, even if $\tau_v = 5s$ for AK5. During the μg phase only one critical parameter exists according to the LSA. The expected pattern is helical. The critical voltages during the μg phase are about $1kV$ lower compared to the critical voltages determined during the $1g$ phase of the PFC. The found flow patterns in the critical parameter range are a mix of helical and columnar pattern. The distribution of these do not seem to follow a certain pattern. The columnar pattern which are found near the experimental parameters for the helical pattern in Fig. 7.25 have often several toroidal vortices near the boundaries. The columnar pattern found in $1g$ or directly at the critical parameter in μg only have one vortex at each boundary. The used LLS/PIV approach only detects $2D$ movements of the particles in a very thin sheet. It could be possible that the boundary effects are misinterpreted parts of a counter-rotating column, which would make these pattern helical. A more sophisticated measurement method, which allows to measure in $3D$, is required to determine the flow pattern. A simpler method could be to use several LLS which are rotated by 45∘ in azimuthal direction.

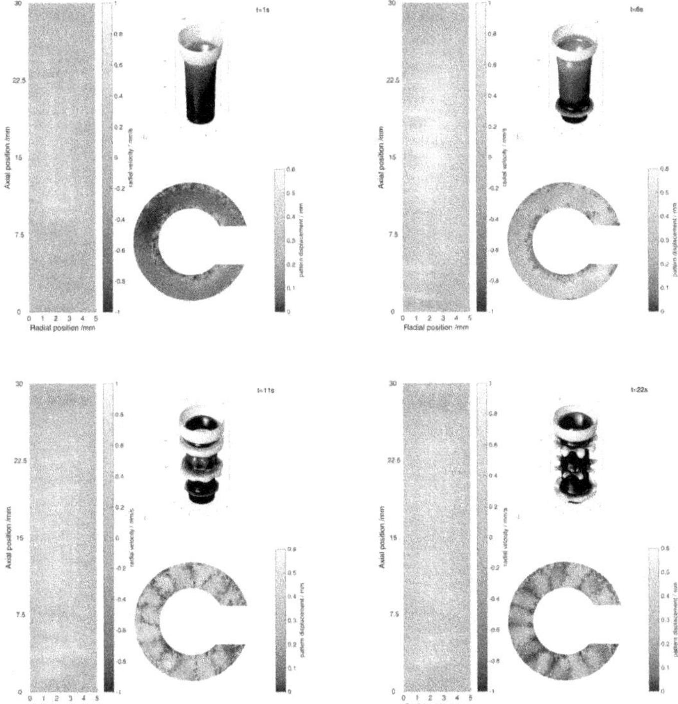

Figure 8.1.: A comparison of the experimental results to the simulations of O. Crumeyrolle (private communication). $t = 1s$ shows the initial state, $t = 6s$ is one quarter into the μg phase, $t = 11s$ is one half into the μg phase and $t = 22s$ shows the situation at the end of the μg phase. The left image shows the radial velocity of the measured flow, the upper right image shows the isothermal surface of the simulated flow and the lower right images shows the evaluated Synthetic Schlieren images of the experiments.

When compared to the 3 simulations of the flow pattern, there is a rather good agreement. The results of Gerstner predict a toroidal pattern in the $100mm$ cell, which can not be confirmed from the experiments. The toroidal pattern is only found in the $30mm$ cell. The helical pattern, which was predicted by Travnikov and Yoshikawa for an annulus of infinite length, can be seen in the $100mm$ cell, but not the $30mm$ cell. The simulation of Crumeyrolle fits the development of the flow pattern in the $30mm$ cell quite well (Fig. 8.1). Right after the start of the μg phase, at $t = 1s$, the experimental images and the the simulation both show a unicellular flow pattern. After $5s$ both, the experiments as well as the simulations show a perturbation. Halfway through the μg phase a toroidal vortex appears at the top in the simulation, which can be seen in the PIV image, and a plume-like structure, which can be seen in the Synthetic Schlieren image. The development of these structures continue and the plume structures can be seen over the whole height of the annulus.

The measurements of the temperature field on the outer cylinder did work out on a global level, but not on a local. It is possible to see the temperature gradient of the unicellular base flow in $1g$ and it also shows a homogeneous temperature distribution in μg , when no voltage is applied. It is also possible to use this temperature data to calculate $\mathcal{N}u$ more precise. However, it does not help to differentiate between the other flow patterns. When there is a columnar pattern, all temperature sensors in the axial direction should give the same temperature, with the exception of boundary effects. The sensors in azimuthal direction should show a gradient with alternating high and low temperatures. This is not the case in either of the used cells. This could be explained by the boundary effects. A slanted pattern should be recognizable by comparing the upper and lower azimuthal measurements. The pattern would be shifted in the upper row when compared to the lower row. I was not able to find this pattern in the temperature measurement. The visualization showed that the azimuthal mode is usually $n = 6$ or $n = 7$. To resolve this in the temperature measurement at least 12 or 14 temperature sensors are required. This would increase the spatial resolution of the temperature measurement and would allows to detect these flow pattern. It would be very challenging to design such an experiment cell. It is also a possibility to use a glass outer cylinder with TCO and additionally a thermochromic coating on the inside. This allows for an area-wide temperature measurement. However, the range of the temperature is limited due to the properties of the thermochromic particles and the compatibility with the high voltage field needs to be tested.

The temperatures measured at the inlet and outlet of the heating and cooling loop are used to calculate the Nusselt number. Although the sensors are outside of the electric field, the signal-to-noise ratio is rather low for small ΔT. This makes the results only reliable for $\Delta T > 10K$ with the used method to calculate $\mathcal{N}u$ (Eq. 3.19). In the previous experiments performed at our facility (Futterer, Dahley, *et al.* 2016 and Dahley 2014) a more simple approach was used, which allowed a higher confidence in the calculation, but also caused many problems with the electric field. Dahley and Futterer used a $100mm$ cell at different voltages and temperature gradients with AK5 in the laboratory. For $\Delta T = 10K$ it was observed that $\mathcal{N}u$ remains constant until $V_p = 5kV$. At $V_p = 6kV$ it started to increase. A similar occurence has been described by Chandra and Smylie (Chandra *et al.* 1972). A similar trend can be found in the data of this thesis. Due to the visualization techniques I can additionally state that this is caused by the change of the flow structure from an unicellular flow to a columnar flow (Fig. 7.24). Unfortunately, the time in the μg phase is too short and it is not possible to reach a stable heat transport. It is only possible to give a rough estimate on how the temperature transport behaves. For both cells $\mathcal{N}u$ is mostly comparable between $1g$ and

μg with the exception of some transition states, where $\mathcal{N}u$ increases by the factor of 2 in the μg phase. This experiments have to be repeated in long-term μg experiments to obtain reliable results.

The experiments with AK5 during the PFC are a good way to examine the behavior of the flow pattern with applied thermal and electric forces. However, to determine the convective heat transfer the time in the μg phase is too short. A TEXUS mission or experiment on the ISS has to be used to determine this. With the work done on the experimental setup and measurement techniques in the scope of this project, a lot of preliminary work has already been completed. But the setup needs to be adapted to be used in TEXUS flights. Another way could be to look at a geometry with a smaller gap. For example τ_k decreases to $\tau_k = 52s$ for a $2mm$ wide gap filled with AK5. When the gap is filled with AK0.65 it decreases to $\tau_k = 47s$. These times are more in line with the restrictions of the parabolic flight. The required voltages would also be lower. However, it is more difficult to visualize the flow in such small geometries. Because of the lower gap width, the structure of the flow may also be different.

Other fluids than AK5 have shown different flow patterns. Some are comparable under certain conditions such as a columnar pattern for AK5 (Fig. 7.46) and AK0.65 (Fig. 7.47). But especially near the inner cylinder it is possible to see oscillatory effects. Due to the lower viscosity it also needs longer to reach a stable flow pattern. This is similar to Novec 7200. Although the observed pattern could also be induced by the particle behavior. No tests have been performed with Novec 7200 and the Shadowgraph or Synthetic Schlieren method. There need to be more long-term tests with these low-viscous fluids in the laboratory to characterize their behavior better.

Another topic which needs clarification is the frequency dependency of the DEP force. The essential assumption is that the Coulomb force can be neglected and only the DEP force has an effect when the frequency of the electric a.c. field is high enough. For AK5 it has been shown that this requirement is fulfilled at $f = 200Hz$. It has also been shown experimentally in this thesis that this frequency is sufficient. This has not been done for the other tested fluids. We only performed experiments to verify the compatibility between the fluid, the tracer particles and the electric field. There was an obvious influence of the electric field on the particles in the first experiments which motivated the need to do more research in this field. Since the focus of this work was not on these fluids the time spend on this experiments was limited. AK0.65 and Novec 7200 have a lower viscosity than AK5. Both show an effect that the particles seem to follow the flow at the outer cylinder but not near the inner cylinder. It is an open question whether this effect is limited to the particles or also to the fluid.

9. Summary and outlook

The influence of the dielctrophoretic force on a dielectric fluid in cylindrical geometry in $1g$ as well as μg conditions has been investigated. The DEP force creates a buoyancy similar to the Archimedean buoyancy. While the Archimedean buoyancy creates fluid flows based on a density gradient and the Earth's gravity, the DEP force creates fluid flows based on a permittivity gradient and an artificial electric gravity g_e. The electric gravity can be used to create fluid flow in a μg environment. Depending on the aspect ratio it is possible to create different flow patterns. The convective heat transfer changes depending on the flow pattern.

One part of the upcoming research should focus on determining the convective heat transfer. It was unfortunately not possible to determine this number with a high confidence due to the limited time in the μg phase during the parabolic flights. For this it is require to do long-term experiments of at least 5 minutes, which is no problem in the laboratory. Experiments in μg require a different approach than the PFC. It needs to be done in a TEXUS mission or on the ISS. A TEXUS mission is already in preparation and the planned execution is in 2021. This should be preceded by long-term laboratory experiments to determine the experimental parameters. It also provides laboratory data to compare the μg data to. This data would allow to calculate a precise $\mathcal{N}u$ to determine the convective heat transfer for each flow pattern.

Another part could concentrate on short-term experiments in PFC. This project is concerned with the flow pattern, but not so much with the determination of $\mathcal{N}u$. It was only possible to find stationary flow pattern in AK5, but not the other used fluids. The flow pattern of AK0.65 or Novec 7200 seemed to be turbulent. The used measurement techniques need to be adapted to verify this. While the combination of 2 2D methods can give an indication of the 3D pattern, a fully 3D visualization method is required. This would make it possible to give a better determination of the flow pattern. It is also possible to use different fluids, such as pure water. The gap could be made smaller, which would reduce the required voltages, but would increase the influence of boundary effects.

A specific application could be the enhancement of the heat regulation system on the ISS or a new station in the future. At the present, this system is divided into two fluid loop. An inner loop, which is filled with water, takes the excessive heat from the rooms in the ISS. The thermal energy is then transferred to the second loop, which is filled with ammonia. The ammonia is circulated through radiators to remove the thermal energy. The temperature of the ammonia is at roughly $-77°C$ and has a permittivity of $\varepsilon > 25$ at this temperature. It would be possible to change this system in a way that the second loop is not circulated by pumps, but by using the TEHD force. The advantage is that no moving parts are required. This reduces the required maintenance. Since ammonia has a high permittivity the required voltage would be low. This could potentially save energy compared to the energy needed by the pumps.

Transferring the acquired knowledge from the μg experiments to an $1g$ application is not feasible based on the short-term experiments. It was not possible to obtain reliable Nusselt numbers and it can not be determined which flow pattern evokes the highest convective heat transfer. The results of the long-term experiments from the TEXUS experiments are required to determine this.

Existing long-term laboratory experiments (Chandra *et al.* 1972; Dahley 2014; Futterer, Dahley, *et al.* 2016; Seelig *et al.* 2019) have shown that the convective heat transfer can be increased in the presented geometry. However, the suggested application in figure 1.3 is not an improvement to the system. The idea is based on a double-pipe heat exchanger, where thermal energy is transferred from a hot fluid to a cooling fluid. The added third pipe, to which the TEHD force is applied, would add a way to control the heat transfer between the other pipes. But it makes the system more complex, does not replace the pumps and does not add any benefits, since the heat transfer can also be controlled by changing the flow rate in the other pipes. The advantages of convections induced by the TEHD force can only be used when pumps are replaced.

The TEHD force can be used to create thermal convection when no gravitational force is present e.g. on a space station, satellites and in other space projects. An application in a $1g$ environment does not seem viable because the geometry needs to be small (several mm) and the applied voltage needs to be high (several kV). Industrial heat exchangers can be several meters big, which would require even higher voltages or special fluids with high permittivity. The main benefit would be to replace pumps with a TEHD-driven convection. This lowers the required maintenance and possibly consumes less energy. However, in a $1g$ application, the TEHD force needs to be strong enough to counter natural gravity. This makes applications in a $1g$ environment less interesting for the geometry used in this thesis.

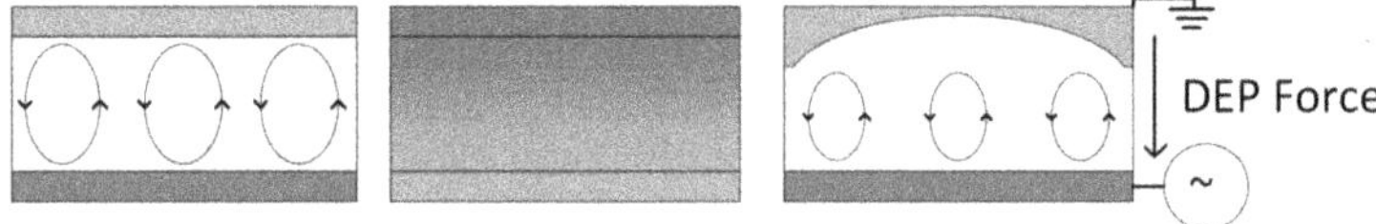

Figure 9.1.: Schematic of the Rayleigh-Bénard convection. A fluid is between two plates. Left: When the lower is heated and upper is cooled it gives raise to the Rayleigh-Bénard convection. Middle: When the upper is heated and the lower cooled a stable temperature gradient establishes. Right: Proposed geometry for TEHD experiments in $1g$. The curvature creates an inhomogeneous electric field, which is directed downward.

Applications in a $1g$ environment should aim at enhancing an existing natural convection. This could involve a Rayleigh-Bénard convection. A fluid is confined in a geometry which is heated from below and cooled from the top. This leads to a natural convection inside the fluid (Fig. 9.1 left). The cold fluid is pulled down by natural gravity while the hot fluid raises upwards due to the volume flow. When the DEP force is applied between the upper and lower plate it could enhance or diminish the convection depending on the geometry. In the case of a stable stratification (Fig. 9.1 middle), when the system is cooled from the bottom and heated from the top, it would be possible to create perturbations. To apply the DEP force efficiently it would require an inhomogeneous electric field. This can be created by introducing a curvature into one of the plates (Fig. 9.1 right). The DEP force would be directed downward since the electric field is stronger at the flat plate than at the curved plate. In this case the natural gravity as well as the artificial gravity would both move the cold fluid towards the bottom plate, which could increase the convective heat transfer.

References

[1] R. J. Adrian, "Particle-imaging techniques for experimental fluid mechanics", *Annu. Rev. Fluid Mech.*, vol. 23, pp. 261–304, 1991.

[2] P. H. G. Allen and T. G. Karayiannis, "Electrohydrodynamic enhancement of heat transfer and fluid flow", *Heat Recovery Systems & CHP*, vol. 15, pp. 389–423, 1995.

[3] A. Bahloul, I. Mutabazi, and A. Ambari, "Codimension 2 points in the flow inside a cylindrical annulus with a radial temperature gradient", *Eur. Phys. J. AP*, vol. 9, pp. 253–264, 2000.

[4] A. Bejan, *Convection Heat Transfer*, 3rd ed. Wiley, 2004, ISBN: 0-471-27150-0.

[5] A. Bergles, *Techniques to enhance heat transfer*, in *Handbook of Heat Transfer*. New York: McGraw-Hill, 1998, ch. 14, ISBN: 978-0070535558.

[6] B. Chandra and D. E. Smylie, "A laboratory model of thermal convection under a central force field", *Geophysical & Astrophysical Fluid Dynamics*, vol. 3, pp. 211–224, 1972. DOI: 10.1080/03091927208236081.

[7] N. Dahley, *Dielectrophoretic flow control of thermal converction on cylindrical geometries*. Göttingen: Cuvillier Verlag, 2014, Dissertation BTU Cottbus-Senftenberg, ISBN: 978-3-95404863-2.

[8] S. B. Dalziel, G. O. Hughes, and B. R. Sutherland, "Whole-field density measurement by 'synthetic schlieren'", *Experiments in Fluids*, vol. 28, pp. 322–335, 2000.

[9] ESA, *ESA User Guide to Low Gravity Platform*, 2014. [Online]. Available: http://www.esa.int/Our_Activities/Human_Spaceflight/Research/European_user_guide_to_low_gravity_platforms (visited on 07/12/2017).

[10] M. T. Fogaing, O. Crumeyrolle, and I. Mutabazi, "Instabilités de convection thermo-électro-hydrodynamiques entre deux parois verticales: Étude numérique", *21éme Congrès Français de Mécanique*, 2013.

[11] B. Futterer, N. Dahley, and C. Egbers, "Thermal electro-hydrodynamic heat transfer augmentation in vertical annuli by the use of dielectrophoretic forces through a.c. electric field", *International Journal of Heat and Mass Transfer*, vol. 93, pp. 144–154, 2016.

[12] B. Futterer, A. Krebs, A. C. Plesa, F. Zaussinger, R. Hollerbach, D. Breuer, and C. Egbers, "Sheet-like and plume-like thermal flow in a spherical convection experiment performed under microgravity", *Journal of Fluid Mechanics*, vol. 735, pp. 647–683, 2013.

[13] T. B. Jones, "Electrohydrodynamically enhanced heat transfer in liquids - a review", *Adv. Heat Transfer*, pp. 107–148, 1978.

[14] M. Jongmanns, M. Meier, A. Meyer, and C. Egbers, "Bestimmung von strömungsmustern in einem zylinderspalt mit radialem elektrischen kraftfeld unter schwerelosigkei", *26. GALA - Fachtagung Experimentelle Strömungsmechanik*, 2018.

[15] M. Jongmanns, A. Meyer, M. M. C. Kang, I. Mutabazi, and C. Egbers, "Experiments on thermoelectric convection in dielectric liquids in a cylindrical annulus under parabolic flight conditions", *International Conference on Rayleigh Bénard Turbulence*, 2018.

[16] L. D. Landau, E. M. Lifshitz, and L. P. Pitaevskii, *Electrodynamics in continuous media*, 2nd ed. Pergamon Press Ltd., 1984, ISBN: 0-08-030276-9.

[17] S. Laohalertdechaa, P. Naphonb, and S. Wongwises, "A review of electrohydrodynamic enhancement of heat transfer", *Renewable and Sustainable Energy Reviews*, vol. 11, 2007. DOI: 10.1016/j.rser.2005.07.002.

[18] S. V. Malik, H. N. Yoshikawa, O. Crumeyrolle, and I. Mutabazi, "Thermo-electrohydrodanamic instabilites in a dielectric liquid under microgravity", *Acta Astronautica*, vol. 81, pp. 563–569, 2012.

[19] O. G. Martynenko and P. P. Khramtsov, *Free-Convective Heat Transfer*. Springer, 2005, ISBN: 3-540-25001-8.

[20] M. Marucho and A. Campo, "Electrohydrodynamic natural convection enhancement for horizontal axisymmetric bodies", *International Journal of Thermal Sciences*, vol. 63, pp. 22–30, 2013.

[21] M. Meier, M. Jongmanns, A. Meyer, T. Seelig, C. Egbers, and I. Mutabazi, "Flow patterns and heat transfer in a cylindrical annulus under 1g and low-g conditions: Experiments", *Microgravity Science and Technology*, 2018, ISSN: 1875-0494. DOI: 10.1007/s12217-018-9649-y.

[22] A. Meyer, *Active control of heat transfer by an electric field*. Université du Havre, 2017, PhD Thesis.

[23] A. Meyer, O. Crumeyrolle, I. Mutabazi, M. Meier, M. Jongmanns, M.-C. Renoult, T. Seelig, and C. Egbers, "Flow patterns and heat transfer in a cylindrical annulus under 1g and low-g condictions: Theory and simulations", *Microgravity Science and Technology*, 2018, ISSN: 1875-0494. DOI: 10.1007/s12217-018-9636-3.

[24] A. Meyer, M. Jongmanns, M. Meier, C. Egbers, and I. Mutabazi, "Thermal convection in a cylindrical annulus under a combined effect of the radial and vertical gravity", *C. R. Mecanique*, vol. 345, 1 2016. DOI: 10.1016/j.crme.2016.10.003.

[25] H. A. Pohl, *Dielectrophoresis*. Cambridge university Press, 1978, ISBN: 0-521-21657-5.

[26] M. Raffel, "Background-oriented schlieren (bos) techniques", *Experiments in Fluids*, vol. 56, 2015. DOI: 10.1007/s00348-015-1927-5.

[27] M. Raffel, C. Willert, S. Wereley, and J. Kompenhans, *Particle Image Velocimetry - A Practical Guide*, 2nd ed. Springer, 2007, ISBN: 978-3-540-72307-3.

[28] H. Richard and M. Raffel, "Principle and applications of the background oriented schlieren (bos) method", *Measurement Science and Technology*, vol. 12, pp. 1576–1585, 2001.

[29] A. Savitzky and M. Golay, "Smoothing and differentiation of data by simplified least squares procedure", *Analytical Chemistry*, vol. 36, pp. 1627–1639, 1964. DOI: 10.1021/ac60214a047.

[30] W. Schöpf, J. C. Patterson, and A. M. H. Brooker, "Evaluation of the shadowgraph method for the convective flow in a side-heated cavity", *Experiments in Fluids*, vol. 21, pp. 331–340, 1996.

[31] T. Seelig, A. Meyer, P. Gerstner, M. Meier, M. Jongmanns, M. Baumann, V. Heuveline, and C. Egbers, "Dielectrophoretic force-driven convection in annular geometry under earth's gravity", *Int. J. Heat and Fluid Flow*, vol. 139, pp. 386–398, 2019.

[32] B. Sitte, *Thermische Konvektion in Zentralkraftfeldern.* VDI Fortschrittsbericht, 2004, vol. 460, Dissertation Universität Bremen.

[33] B. Sitte and H. J. Rath, "Influence of the dielectrophoretic force on thermal convection", *Experiments in Fluids*, vol. 34, pp. 24–27, 2003. DOI: 10.1007/s00348-002-0524-6.

[34] P. J. Stiles, F. Lin, and P. J. Blennerhassett, "Convectice heat transfer through polarized dielectric liquids", *Phys. Fluids A*, vol. 5, 12 1993. DOI: 0899-8213/93/5(12)/3273/7/$6.00.

[35] M. Takashima, "Electrohydrodynamic instability in a dielectric fluid between two coaxial cylinders", *Q. J. Mech. Appl. Math*, vol. 33, pp. 93–103, 1 1980.

[36] W. Thielicke and E. Stamhuis, "Pivlab – towards user-friendly, affordable and accurate digital particle image velocimetry in matlab", *Journal of Open Research Software*, vol. 2, 2014. DOI: http://dx.doi.org/10.5334/jors.bl.

[37] V. Travnikov, O. Crumeyrolle, and I. Mutabazi, "Numerical investigation of the heat transfer in cylindrical annulus with a dielectric fluid under microgravity", *Physics of Fluids*, vol. 27, 2015. DOI: 10.1063/1.4921156.

[38] R. J. Turnbull, "Effect of dielectrophoretic forces on the bénard instability", *Physics of Fluids*, vol. 12, 1969. DOI: 10.1063/1.1692745.

[39] H. Yoshikawa, O. Crumeyrolle, and I. Mutabazi, "Dielectrophoretic force-driven thermal convection in annular geometry", *Physics of Fluids*, vol. 25, 2013. DOI: 10.1063/1.4792833.

A. Measurement hardware

A.1. Visualization hardware

Depending on the experiment box different cameras are used.

Table A.1.: Technical parameters of the cameras. The IDS cameras are used in the T Box for the Shadowgraph and Synthetic Schlieren methods. The ImagingSource cameras are used in the P Box.

	IDS UI-5550SE	The ImagingSource DMK33GX174
Resolution	1600x1200 pixel	1920x1200 pixel
Color Depth	12 bit	12 bit
Sensor type	CMOS	CMOS
Actual framerate	10 FPS	10 FPS
Shutter type	rolling	global
Objective	Edmund Optics 59-872 with focal length $FL = 35mm$	The Imaging Source FL-HC1214-2M with focal length $FL = 12mm$

The lighting in Experiment Box T is realized by a LED-array (CCS TH-51X51RD) with red light ($\lambda = 635nm$) and a maximum power of $5W$. The actual used power is lower and the intensity be changed by a potentiometer. It depends how many light control films (LCF) and masks are used. The setting has to be re-adjusted when the measurement method is changed from Shadowgraph to Synthetic Schlieren method. The LCF are used to create quasi-telecentric light. One vertical (CCS LC-TH-51X51-VE) and one horizontal (CCS LC-TH-51X51-HO) film are used per cell. The quality of the resulting telecentricity is not as good as with a more dedicated optical setup, but it is very compact and requires a lot less space inside the Experiment Box.

The laser used inside Experiment Box P is a MediaLas LLM-100-650 line laser. It is a continuous line laser with red light ($\lambda = 620nm$) and power of $100mW$.

Table A.2.: Properties of the PIV particles per fluid.

Fluid	Particles	$\rho/g/cm^3$	ε
AK5	Potters Hollow Glass Spheres	1.1	4.8
AK0.65	Potters Sphericels 60P18	0.6	5.8
Novec 7200	Polyamide P84-NT	0.6	3

Functional tests of the cameras

I found possibly oscillatory structures in the Shadowgraph images captured with the IDS cameras. It was not clear whether this effect was an actual flow inside the oil or somehow caused by the cameras. This effect was not very apparent in the PFC results due to the limited time frame. The oscillations occur every $10s$, which is about half of the μg time. But in longer experimental runs they become more prominent. To verify this I performed an experiment to resolve this issue.

To do this I placed the cameras in front of a white sheet of paper. For the measurements all automated functions of the cameras, such as auto focus, white balancing, etc. were deactivated. A video of 1 minute of the white sheet was captured at $10Hz$. It would be expected to see maybe some noise but no other effects, since no settings of the camera or the lighting changes during this time. The images are visualized similar to the Shadowgraph method, so that these images can be compared. The average image over the 1 minute sequence is calculated and then normalized by the first image of the sequence. The IDS camera shows a striped pattern where the light intensity slightly increases and decreases (Fig. A.1). When one looks at the time dependent change (Fig. A.2) it is apparent that this causes the oscillatory pattern in the measurements. Since the camera uses a rolling shutter I assume that the pattern we can see is caused by the shutter. The ImagingSource cameras, which have a global shutter, do not show this pattern, but only some evenly distributed noise (Fig. A.3).

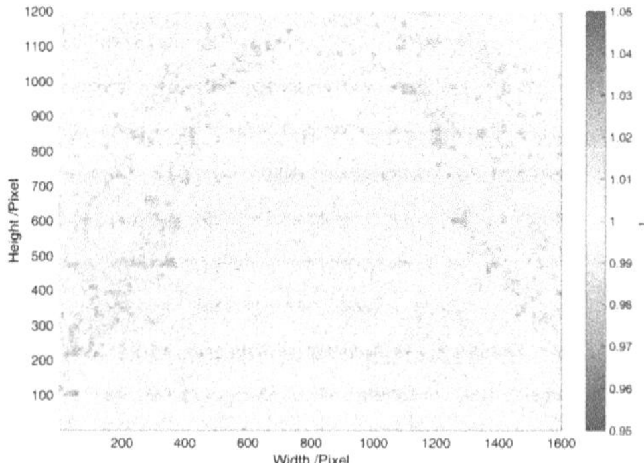

Figure A.1.: The IDS cameras recorded a white background and the averaged image was normalized by the first image. The effect of the rolling shutter is visible.

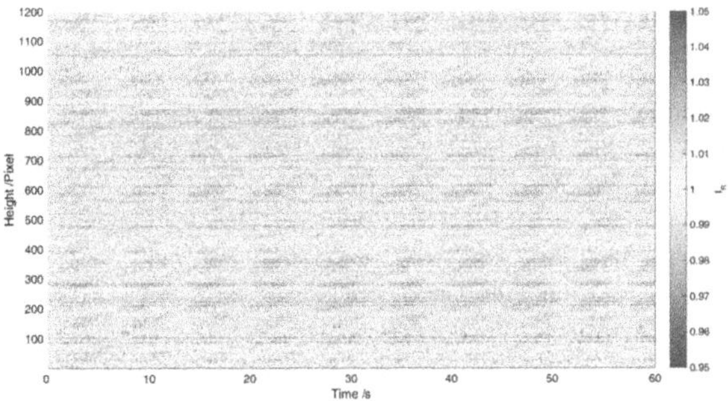

Figure A.2.: A space-time plot of the sequence captured by the IDS cameras. The white background does not change, but the acquired light level changes periodically.

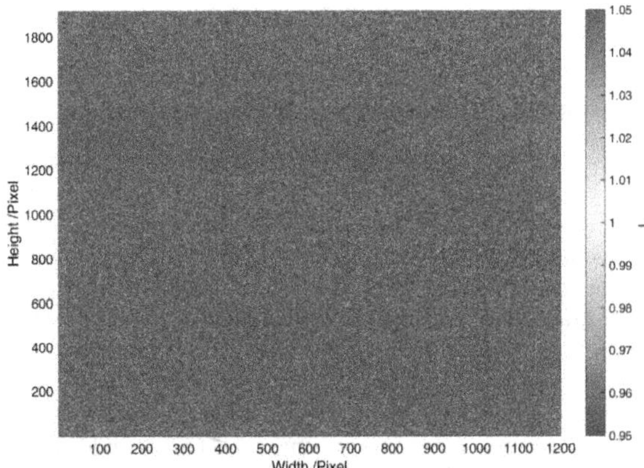

Figure A.3.: The ImagingSource cameras recorded a white background and the averaged image was normalized by the first image. Some evenly distributed noise is visible.

A.2. Temperature measurement

The thermocouples used for the type T cells are type J with a very small diameter of $0.08mm$ (OMEGA 5TC-TT-JI-40-1M). They are used for measurements on the outer cylinder. Since they are very small they can be fitted easily inside the cylinder wall and react fast to temperature changes in the fluid.

The used PT100 sensors are RS Pro PT100, $2.3mm$, class A. They are used inside the heating and cooling loops.

The following table gives an overview over the used temperature measurement converters.

Table A.3.: Technical parameters of the measurement converters. Depending on the used type of thermocouple and required accuracy, different converters are used.

	Seneca K109TC	Seneca K109PT	OMEGA TXDIN70
Accuracy	0.1%	0.1%	0.3%
Temperature drift	$120ppm/K$	$100ppm/K$	$\leq$ 0.015% of end value/K
Set measurement range	$0\ldots200°C$	$0\ldots100°C$	$0\ldots100°C$
Response time	$< 55ms$	$< 200ms$	
Used Thermosensor	Thermocouple type J	PT100	Thermocouple type J

Influence of the electric field on the measurements

Due to the strong electric fields inside the Experiment Box it can be expected that the EMI has an impact on the temperature measurements. Not only does the amplitude of the electromagnetic field influence the signal of the thermocouples, but also the frequency can cause resonance-based effects. To investigate this effects I used temperature measurement acquired at different frequencies of the voltage. In the first experimental run the temperature gradient is fixed at $\Delta T = 10K$ and the voltage at $V_p = 10kV$ (Fig. A.4). The frequency of the voltage was in the range of $0.1Hz$ to $200Hz$ with an interval of $20Hz$. The mean and the standard deviation was calculated for each frequency over a data set of 5 minutes. It is possible to see that the standard deviation changes depending on the temperature. It has a minimum of $\pm0.1K$ around $0.1Hz$, $100Hz$ and $200Hz$. There is a maximum deviation of nearly $\pm0.3K$ at around $40Hz$ and $140Hz$. This could be caused by a direct effect of the electromagnetic field on the temperature measurement system or indirectly by an effect of the PID controller, which regulates the temperature.

To eliminate the possibility that these effects are caused by the PID controller I did a second experimental run. The only difference to the previous one is that the heating power is set to a constant value (Fig. A.5). The PID controller is deactivated and has no effect in this situation. The temperature was not yet very stable during the first two measurements at $0.1Hz$ and $20Hz$, but become more stable starting at $40Hz$. The minima and maxima of the standard deviation are comparable to the previous situation where the PID controller is active. Hence, this effect cannot be caused by the PID controller, but by an interaction of the electric field on the temperature measurement hardware.

An excerpt of the measured temperature and voltage data at $f = 0.1Hz$ is given in figure A.6. In this run the following parameters are used $\Delta T = 10K$, $V_p = 10kV$ and $f = 0.1Hz$. The curves of the temperature measurements seem to follow the voltage. Unfortunately it is not possible to make this plot with higher frequencies, because the sampling rate of the data acquisition system is too low.

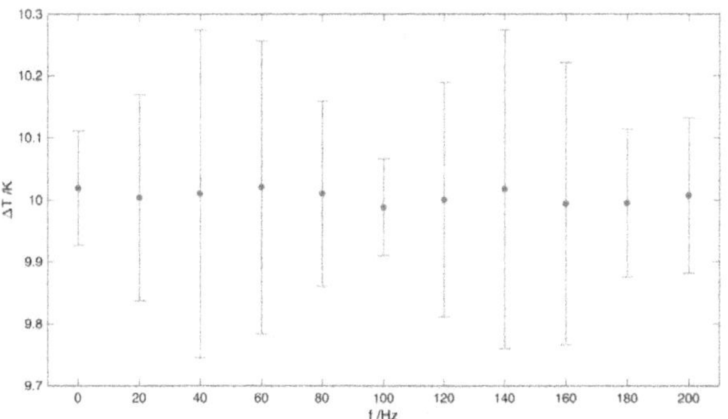

Figure A.4.: The temperature gradient is regulated by the PID controller to $\Delta T = 10K$. The voltage is activated at $V_p = 10kV$ with a variable frequency of $f = 0.1Hz$ to $f = 200Hz$. The mean temperature is always close to $10K$, but the standard deviation shows a correlation to the frequency.

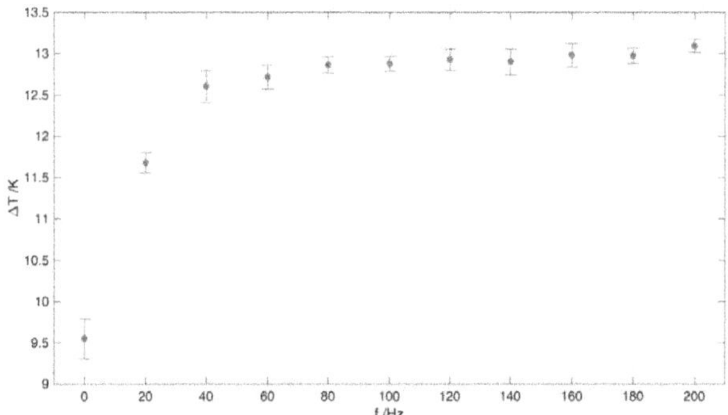

Figure A.5.: The heating power is set to a fixed value and no regulation acts on the heating loops. The voltage is activated at $V_p = 10kV$ with a variable frequency of $f = 0.1Hz$ to $f = 200Hz$. The mean temperature shows an increase over time and the standard deviation shows a correlation to the frequency.

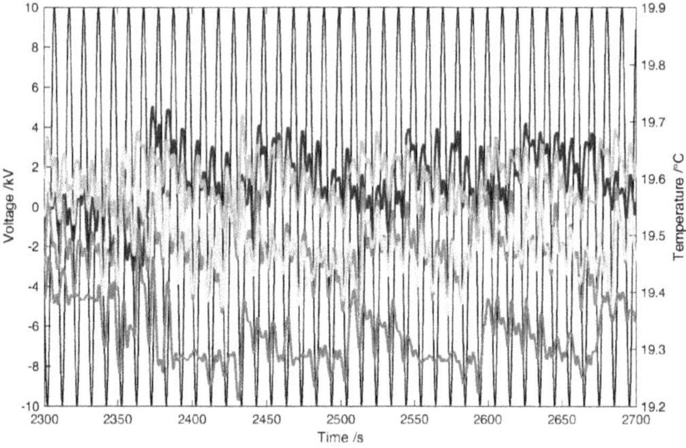

Figure A.6.: The voltage curve of $0.1Hz$ is shown (black) in comparison to the measured temperatures of 5 thermocouples inside the wall of the cell. Whenever the voltage reaches a maximum there is also a maximum in the temperatures.

A.3. National Instruments components

Two data acquisition cards from National Instruments are used to acquire the needed data.

Table A.4.: Technical parameters of the National Instruments DAQ cards.

	PCI-6255	PCI-6259
ADC		
Number of channels	80	32
Resolution	16 bit	
Absolute accuracy	$410\mu V$	
Sensitivity	$22.8\mu V$	
Random noise	$57\mu V_{rms}$	
Temperature offset	$24ppm/K$	
DAC		
Number of channels	2	4
Resolution	16 bit	
Absolute accuracy	$2mV$	
Temperature offset	$17ppm/K$	

The early experiments were performed with LabVIEW version 2012. This was later updated to LabVIEW 2017 64-bit (version 17.0.0) with the NI Vision Acquision Software 2017 (IMAQ version 17.0.0) and IMAQdx 17.1.0.

A.4. Assorted components

The PC used to control the experiment is running Windows 7 SP1 64-bit version on an Intel i7-3770K 3.5GHz processor with 8GB RAM. The data is saved on a SSD and later transferred to our department server for evaluation.

The high-voltage amplifier is a TREK 10/10B.

Table A.5.: Technical parameters of the high voltage amplifier. Depending on the used experiment cells a sine wave of up to $500Hz$ with an amplitude of $10kV$ can be generated.

Output voltage	$-10\ldots+10kV$
Output current	$-10\ldots+10mA$
Slew rate	$> 700V/\mu s$
Output noise	$< 0.5V_{rms}$

The flow rate inside the heating fluid loop is measured with the mass flowmeter Biotech FCH-m-POM. It generates a pulse on the output when a certain volume has passed. This has been calibrated using AK5, which is used as heating fluid, with an accuracy of $\pm 0.33ml/s$.

To determine the acceleration forces acting on the experiment, an accelerometer is mounted on the Control Rack and connected to the PC via USB. It is a Code Mercenaries CM JW24F14. We use 14-bit resolution per axis with a measurement range of $\pm 3g$.

A.5. Fluid properties

The following liquids have been used during the experiments. The values of the properties are either taken from the respective datasheets of the manufacturers or measured in our laboratory. The temperature dependent density of AK5 has been determined as $\rho(T) = -0.9\frac{kg}{m^3 \cdot {}^\circ C} * T + 937.8\frac{kg}{m^3}$, where T is the fluid temperature in $^\circ C$.

Table A.6.: Physical parameters of the used experimental fluids.

	Unit	AK5	AK0.65	Novec 7200
Density	ρ in kg/m^3	920	760	1430
Kinematic viscosity	ν in m^2/s	$5 \cdot 10^{-6}$	$0.65 \cdot 10^{-6}$	$0.43 \cdot 10^{-6}$
Relative permitivity	ε_r	2.7	2.18	7.3
Electric conductivity	σ in S	$1 \cdot 10^{-12}$	$1 \cdot 10^{-12}$	$1.46 \cdot 10^{-11}$
Prandtl number	$\mathcal{P}r$	64.6	7.7	11
Thermal conductivity	λ in $W/K \cdot m$	0.12	0.1	0.07
Thermal expansion	α in K^{-1}	$1.08 \cdot 10^{-3}$	$1.34 \cdot 10^{-3}$	$1.6 \cdot 10^{-3}$
Thermal diffusivity	κ in m^2/s	$7.74 \cdot 10^{-8}$	$8.49 \cdot 10^{-8}$	$3.90 \cdot 10^{-8}$
Specific heat capacity	c_p in $J/kg \cdot K$	1630	-	-
Thermal relaxation time	τ_k in s	323	294	641
Viscous relaxation time	τ_v in s	5	38	58
Electric relaxation time	τ_e in s	23.9	19.3	2.6

B. LabVIEW program

The experiment is controlled by a National Instruments LabVIEW program. The first version of the interface was developed in 2012 by a student. I reworked the program completely and continued to improve it for each PFC. In the current state the program is completely automated and can be synchronized to the gravity phases of a PFC. All variables can be set via a scripting system and the scripts for the PFC can be tested with a PFC simulation, which uses the gravity data from previous flights. It is, of course, also possible to set all parameters manually.

B.1. User interface

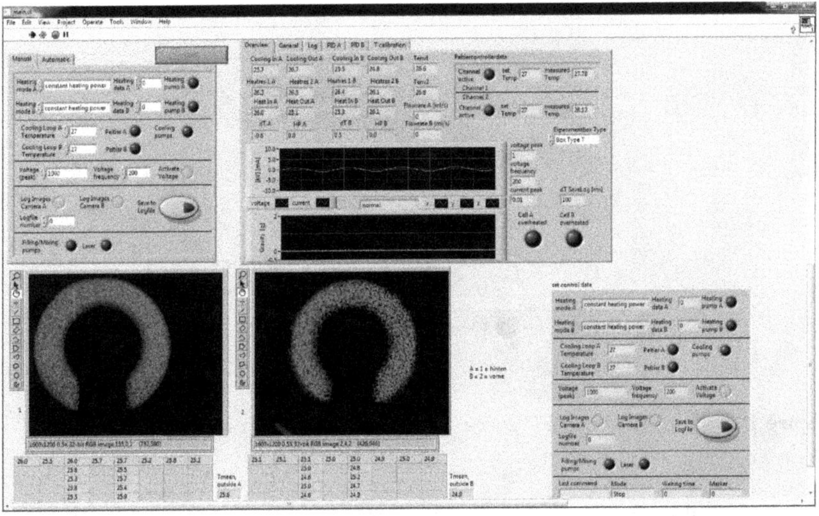

Figure B.1.: LabVIEW main interface (Version 2017) during an experimental run using the Synthetic Schlieren method.

In the top left corner of the interface (Fig. B.1) it is possible to choose between a manual and automatic mode. In manual mode all aspects of the experiment can be set manually, while in automatic mode a script file can be loaded to perform automated experiment runs. The *set control data* panel displays the actual set data. The program performs a plausibility check for the set data, if e.g. the peak voltage is set to 100kV, then the actual set data will be limited to 10kV, since this is the maximum of the high voltage amplifier. In the lower left the image and temperature data from the outer cylinder of the cells is displayed. In the upper right an overview of the control temperatures is given. This includes heating loop and cooling loop temperatures, the actual temperature gradient, the gravity data from the accelerometer and the feedback from the high voltage amplifier.

Control mode

The program can be operated in a manual or automatic control mode.

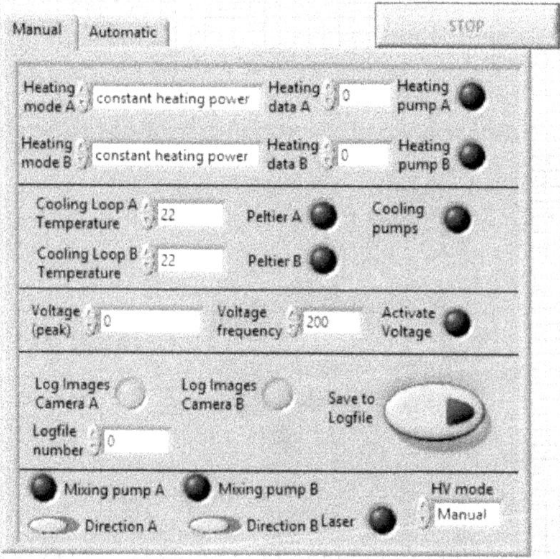

Figure B.2.: LabVIEW panel (Version 2018.2) for manual control of the experiment.

In the manual mode (Fig. B.2) it is possible to set all data manually. The heating systems and pumps of the heating loop can be independently. When the hating mode is set to *constant heating power*, the heating cartridges are controlled at a fixed percentage of their maximum power. The heating data is then the percentage and can be in a range of $0 \ldots 100$. When it is set to *constant temperature difference*, the heating power is controlled by a PID regulator. The heating data is the desired ΔT between the heating and cooling loops. The pumps should be always active when the heating is active, but can also be controlled manually which gives more flexibility for certain experiment settings. The temperature of the cooling loop is set to a constant temperature in $°C$ and controlled by the peltier controller. This controller can run independently from the LabVIEW program and communicates via a serial interface with LabVIEW. The voltage is given by the peak voltage V_p in V and the frequency in f. The output signal is calculated by $y(t) = V_p \cdot sin(2\pi ft)$. When *Save to Logfile* is activated, all measured data and images are written to the hard drive. After it is deactivated, the *logfile number* is increased by one. This way each experimental run has its own number and can be identified easier. It is also possible to disable the logging of the image data when only the temperature data is of interest to safe disc space. The mixing pumps can be activated independently for each cell to disperse the particles. It is also possible to change the direction in which the fluid is pumped

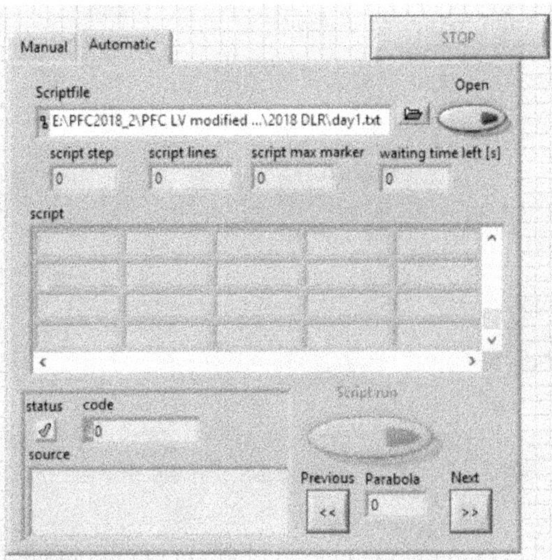

Figure B.3.: LabVIEW panel (Version 2018.2) for automatic control of the experiment.

In the automatic mode (Fig. B.3) a script can be loaded, which then controls the complete experiment flow. First a *script file* needs to be selected and then loaded with the *open* button. This reads the file and performs an error check. When no errors were found, the *script run* button becomes active. Otherwise the type of error and line in which this error was found is displayed in the error section. After a click on *script run*, the script is started and the LabVIEW program takes control of all parameters. Within the script it is possible to define markers. During the PFC these markers correspond to the number of the parabola. With the *previous* and *next* buttons it is possible to jump to the previous or next marker position. This allows to synchronize the script to the current parabola if the automatic parabola detection should fail.

Data display

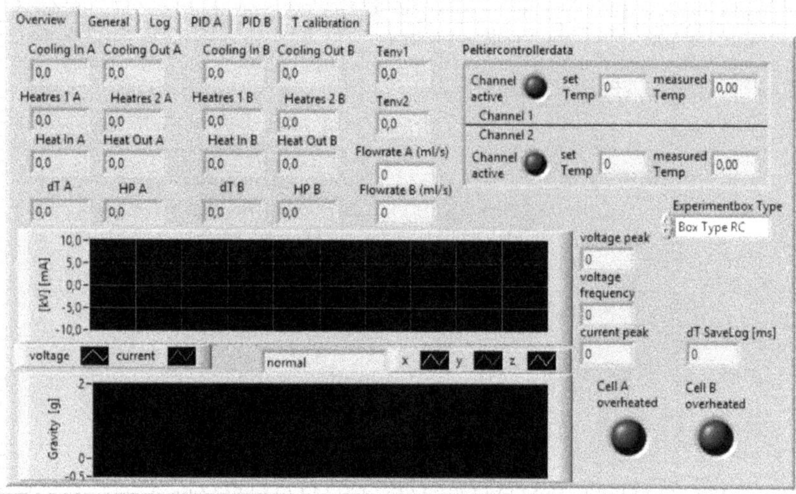

Figure B.4.: Overview panel (Version 2018.2) for experimental data.

An overview of the core data and feedback of the experiment is shown in Fig. B.4. It shows the temperatures of the inlets and outlets of the fluid loops. The temperature at the inlet of the cooling loop is used to control the targeted temperature of the fluid loop. Similar, the temperature of the inlet of the heating loop is used to control the desired temperature difference. The *heatres* temperatures show the temperatures of the heating reservoir. If these temperatures exceed a certain limit, the heating cartridges are deactivated. This prevents an overheating of the fluid. If this happens, the big indicators in the bottom right corner will light up red. The top right area shows the data acquired from the peltier controller. If this data does not change every second, there is a communication failure between the controller and the LabVIEW program. The upper graph shows the feedback data of the high voltage amplifier. This way it is possible to see whether the set peak voltage and frequency is reached or not. The lower graph shows the data acquired from the accelerometer.

Settings

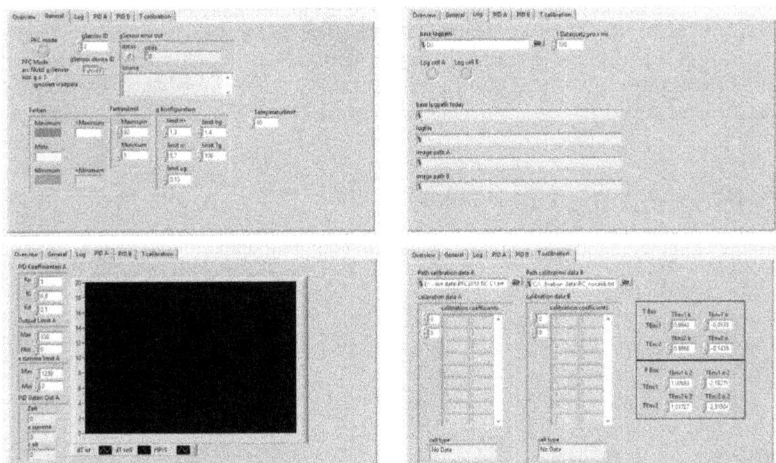

Figure B.5.: Panels to change the settings of the LabVIEW program (Version 2017).

Several settings can be made to the program directly in the user interface. In the *General* tab (Fig. B.5 top left) it is possible to configure the accelerometer and several thresholds. The accelerometer is active when *PFC mode* is activated. The associated ID is determined by windows and has to be set manually to the correct number. When the *PFC mode* is deactivated, the g_z data is acquired from the PFC simulation, which allows to test the scripts for parabolic flights on the ground. The colors correspond to the upper, middle and lower temperatures. These colors and limits determine how the background color of the temperature data is calculated. The *g configuration* includes the thresholds for determining the current phase of the parabolic flights. When $g_z <$ limitμg then microgravity is reached. When $g_z >$ limit hg then hypergravity is reached. When limit n- $< g_z <$ limit n+ for limit $1g$ ticks then the a steady flight is reached. This system operates at 10 ticks per second. The *temperature limit* gives the maximum temperature of the heating fluid loops. If the temperature of any sensor inside the heating fluid loop exceeds this limit, then the heating cartridge is deactivated.

The *log* tab (Fig. B.5 top right) allows to change the base path where the log files are saved. The number next to the path determines at which interval in ms a data set is recorded. The actual paths are determined automatically and also include the date and time at which the LabVIEW program was started.

Each heating loop has its own tab to set the parameters of the *PID* regulator (Fig. B.5 bottom left). The parameters k_p, k_i and k_d are the scaling parameters for the proportional, integral and derivative parts respectively. The *output limit* is set to $0 \dots 100$ which corresponds to an heating power of $0\% \dots 100\%$. The error sum of the integral part is limited as well to increase the response time of the output, when the input changes.

The *T calibration* tab allows to include calibration factors for the temperature sensors. The calibration is performed as described in chapter 6.1. The data format of the files is a simple tab-delimited format. The first value is the coefficient c and the second value the offset o. The actual temperature is then determined by $T_{calibrated} = T_{measured} * k + o$.

B.2. Program flow

The program flow is split into several timed loops (Fig. B.6). This ensures that the acquisition and control of data can be performed independently from each other. LabVIEW executes each loop as its own thread. These threads are controlled by either a timer or event system. The order in which these threads are executed is determined by LabVIEW and the timing is usually precise to $1ms$. An overview of these individual loops is shown in the block navigation view (Fig. B.6). The individual sections are described in this chapter.

Section A controls the connection to the peltier controller. In the first frame of the sequence, the NI-VISA driver to control the serial interface is initialized. The connection between LabVIEW and the controller is checked continuously. When the check fails, all drivers are reset and the connection is re-initialized. Then the communication is synchronized. The settings for the controller are send by LabVIEW after a handshake sequence. These settings are saved in the *peltier controller* section of LabVIEW. Then the desired and measured temperatures are exchanged every second.

Section B acquires the feedback data of the high-voltage amplifier. One section reads the voltage and calculates the peak voltage and frequency of the signal. The other section reads the feedback of the current and calculates the peak current. Both data acquisitions run without temporal restrictions, which results in near real-time data.

Section C controls the output of the DAQ hardware. The inputs made in the UI are send to the corresponding physical output channels. The digital outputs are mapped directly to their channel. The output for the high-voltage amplifier and heating cartridges are generated by the signal generation function integrated in LabVIEW. The signal for the high-voltage amplifier is a sinus curve. The heating cartridges are controlled by a pulse-width modulated signal. The frequency of this signal is fixed at $20Hz$ and the time this impulse is high is controlled by the program. By modulating the length of the signal it is possible to change the effective heating power. At $20Hz$ each part of the signal has a duration of $50ms$. When the signal has a high level for $25ms$ and low level for $25ms$, then the effective heating power is at 50%. Similar, a high level time of $5ms$ and low level time of $45ms$ corresponds to an effective heating power is 10%. When an overheating of the fluid loop is detected, the output signal is forced to a low level. Both, the signal for the high-voltage amplifier and the heating cartridges, are mapped to an DAC channel of the DAQ hardware.

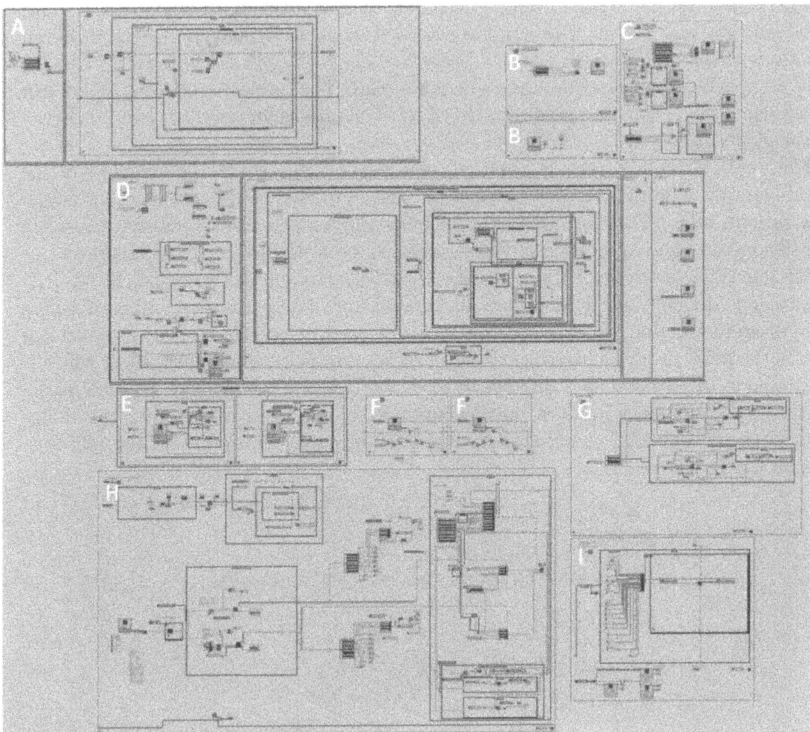

Figure B.6.: The LabVIEW block navigation view shows the functional units of the program.

The 3-frame sequence in section D controls the main flow of the program. The first frame initializes all variables and loads the temperature calibration data automatically, if this function is activated. The second frame is the main loop of the program which either transfers the desired experimental parameters to the set parameters or executes the script in automatic mode. The third frame stops all DAQ processes and deactivates all physical outputs. The main function in the second frame also includes an event control. It reacts to button presses in the UI to e.g. load and verify the script data or change the position in the script. When no event is detected the experimental flow is actualized. In manual mode, the desired parameters are limited to the operational range of the experiment, e.g. the peak voltage cannot exceed $10kV$, and then set as actual output parameters. In automatic mode the script is executed. This includes a combination of event driven and time driven commands. The execution of the script can be paused until a certain event is detected, such as reaching the μg phase, or until a certain time has passed. The script is executed sequentially and can (de-)activate all functions of the program.

In section E the images from the camera are read. The framerate of each camera can be set using the vision acquisition wizard of NI. The images are acquired and also saved at the given rate.

The data of the flowmeter is acquired in section F. The number of impulses of the flowmeters are acquired over the course of $1s$. These are then used to calculate the actual flow rate im ml/s. The calculation also includes the temperature dependent density of the silicone oil. The calibration of this system has been performed in our laboratory.

Section G includes the PID regulators used to maintain the desired ΔT. A PID regulator is divided into a proportional, integral and differential part and realized as time discrete function. In this program it is executed every $100ms$. The proportional part is the basic response based on the error. The error is integrated over time, which is then used in the integral part, and allows to create a long-term stability. The differential part uses the difference between the current and previous error, which gives the ability to react fast to sudden changes. The regulator is implemented like in the following pseudocode:

$$e = T_{target} - T_{measured}$$
$$e_{sum} = e_{sum} + e$$
$$e_{sum} = \text{limit}(e_{sum}, \text{minimal\_value}, \text{maximal\_value})$$
$$y = k_p \cdot e + k_i \cdot e_{sum} \cdot t + k_d \cdot \frac{e - e_{previous}}{t}$$
$$y = \text{limit}(y, 0, 100)$$
$$e_{previous} = e$$

with the coefficients k_p, k_i and k_d to determine the controller behavior, t the time in seconds between the current and last execution of the code, T_{target} the target temperature or temperature difference, $T_{measure}$ the current measured temperature or temperature difference and y the controller output, which is passed on to the heating control. Before passing the control value to the control of the heating system it is limited to the range of 0...100, which corresponds to a heating power of 0...100%. The error sum is also limited by a minimum and maximum value, which is set in the *PID settings* tab.

Section H acquires the data from the sensors and writes one data set to the current log file. The time step at which this is called can be set in the *log settings* tab. One part of this loop fetches the data from the accelerometer. This data is passed to the system which keeps track of the current phase of the parabolic flight. This data is then used by the automatic mode in section D. The data from the temperature sensors are also acquired. There are 60 analog channels, which are only utilized completely by Experiment Box type T. The other Experiment Boxes only use about 20 sensors. Each Experiment Box has a different mapping from the physical analog channels to the position of the sensors, e.g. heating loop inlet or outlet. The calibration data are applied here as well. Then the data is displayed at the corresponding part of the UI. When data logging is active, the data is written to the log file. The files are tab-delimited text files, where each line is one data set. Each line contains a timestamp, accelerometer data, temperature data,feedback of the high-voltage generator, image number, etc. Since the image acquisition and data acquisition is not synchronized, it is possible to e.g. capture the images at $20FPS$, while the temperature data is saved at $10Hz$. The number of the current image is written to the log file to allow for a synchronization of both data during the evaluation.

Lastly, section I is optional, but adds to the human readability of the data, especially when using Experiment Box type T. It scans all temperature data and maps a color to the corresponding UI element. This is done at a low priority and takes $1s$ to update all data fields. This way it is easier to recognize when a temperature, such as in the heating loop, exceeds a certain limit or when the connection to a thermoelement is damaged.

Flowcharts

Since LabVIEW uses a graphical programming approach it already is a flowchart in itself. I show only a few additional charts to explain the basic idea behind the design.

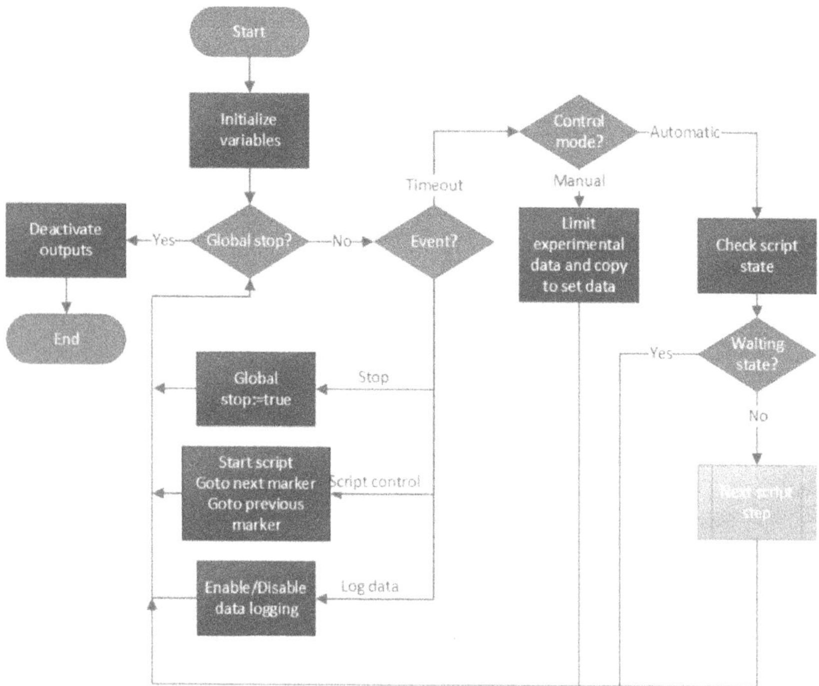

Figure B.7.: Flowchart of the main part of the LabVIEW program. The event system included in LabVIEW reacts to events like button presses and returns a timeout when no events are detected.

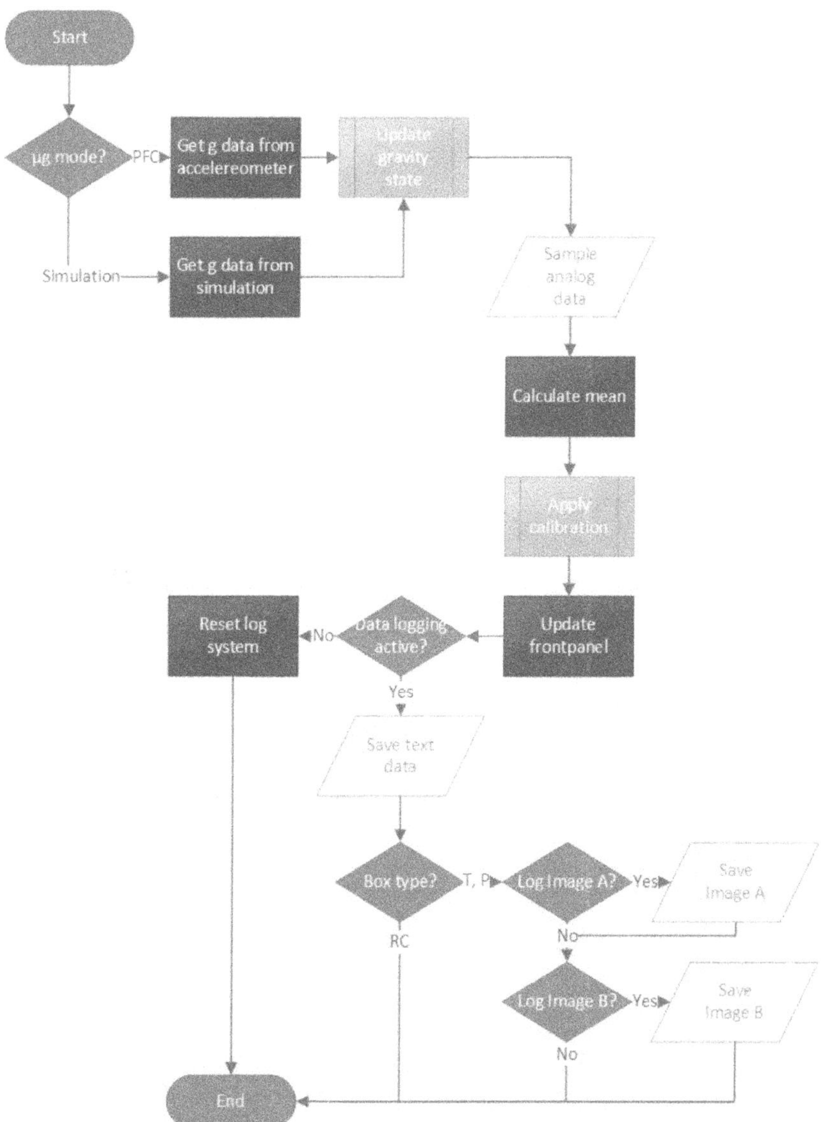

Figure B.8.: Flowchart of the part which handles the data acquisition. The acquisition of the image data for Box type RC is asynchronous from the temperature data and is done in section E (Fig. B.6).

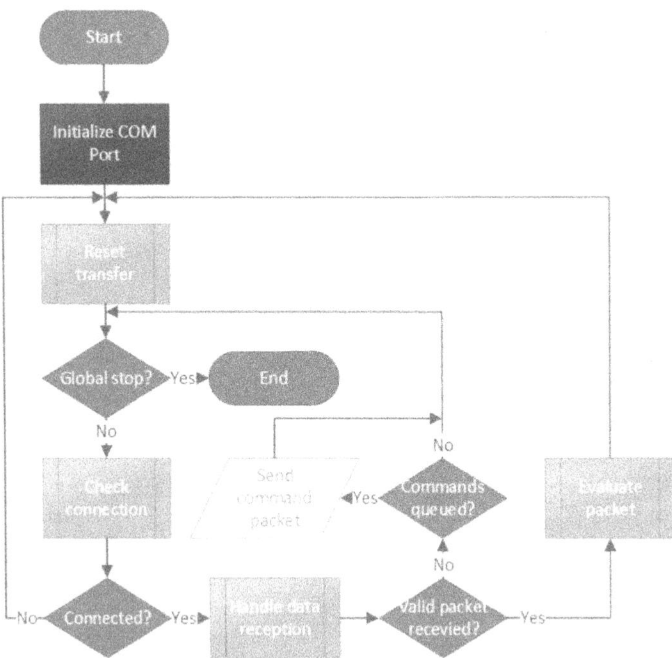

Figure B.9.: Flowchart of the part which controls and monitors the connection to the Peltier controller. Reset transfer clears all variables and sets the state to the beginning of a transfer.

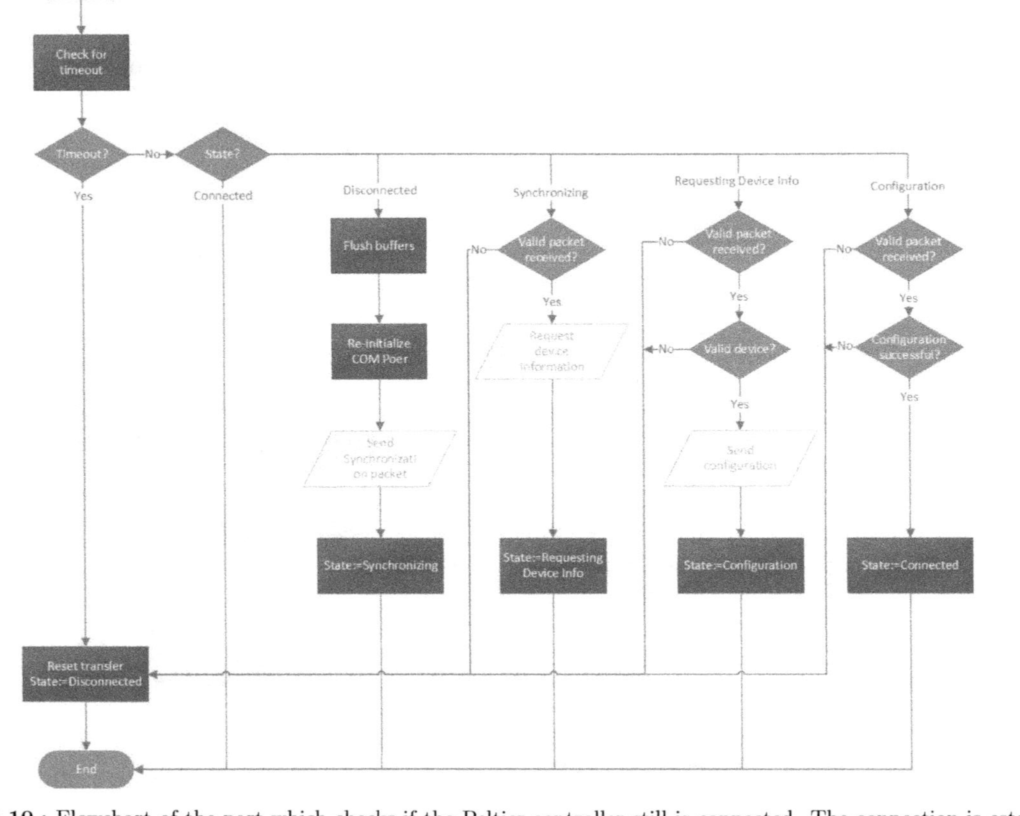

Figure B.10.: Flowchart of the part which checks if the Peltier controller still is connected. The connection is established step-wise. When successful this subprogram also sends the configuration data.

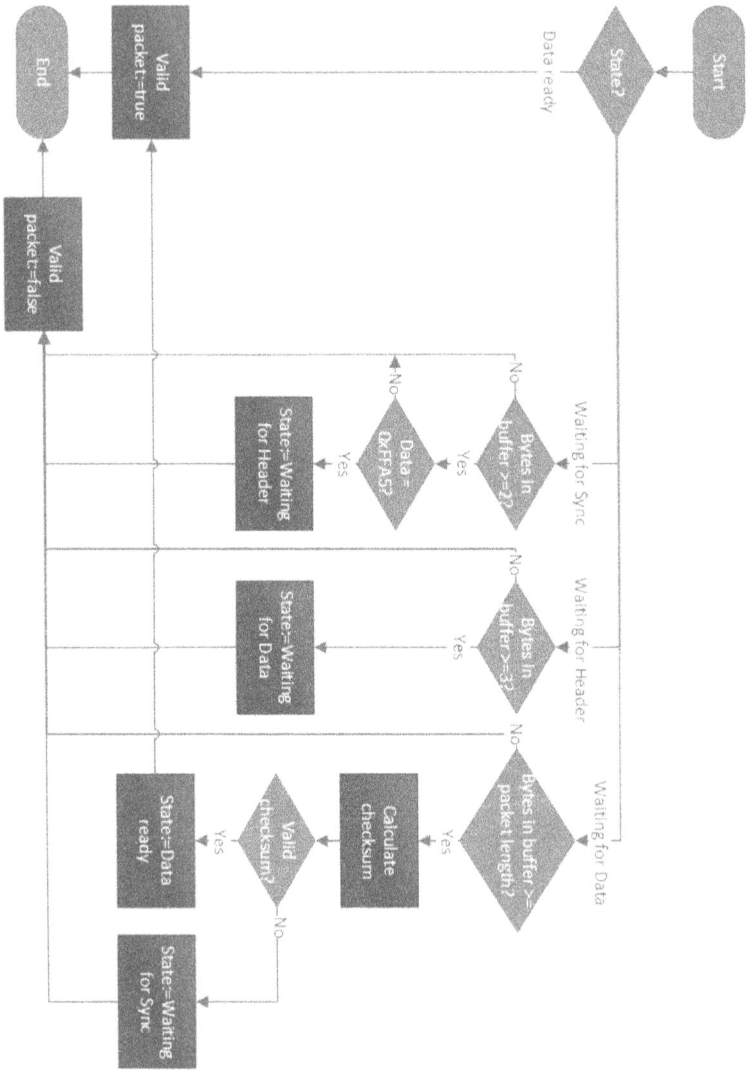

Figure B.11.: Flowchart of the part which controls the data flow. The data is evaluated as part of the main program, which also handles the timeout and resets the process after the packet has been evaluated.

B.3. Scripting interface

The script is a text file with one command per line. Lines starting with # are comments and ignored by the script parser. The available functions and commands are explained for the latest versions of the scripting system.

The Syntax is given as **command** *parameters*. Neither of these are case-sensitive. The *parameters* are given as expected parameter type and the default value. The possible parameter types are:

num A numeric value. The function depends on the command.
str An alpha-numerical value. The function depends on the command.
cell Can be *a* or *b*. Determines the cell for which the command is valid.
oo Can be *on* to activate a feature or *off* to deactivate a feature.

The *cell* parameter needs always to be specified. The *oo* parameter defaults to *off* if not specified. This prevents accidental activation of components. If the *num* parameter is not given it can default to different values depending on the command, which is indicated by *=value* in the parameter list.

Set peak voltage

Syntax: **setV** num=0

The high voltage is applied as sinewave in the form of $A * sin(2\pi ft)$. This command sets the amplitude A, which represents the peak voltage of the signal. Setting $num = 0$ does not automatically deactivate the output from the amplifier and setting $num \neq 0$ does not automatically activate it. For safety reasons the voltage should be deactivated using the voltage command instead of setting it to 0.

Set voltage frequency

Syntax: **setVf** $num = 200$

The high voltage is applied as sinewave in the form of $A * sin(2\pi ft)$. This command sets the frequency f.

Activation of the voltage

Syntax: **voltage** $oo =$ off

This can be used to activate (oo=on) or deactivate (oo=off) the output of the high voltage amplifier.

Set temperature gradient

Syntax: **setT** *cell* $num = 0$

This command activates the PID regulator to generate a constant temperature gradient for the given cell with $\Delta T = numK$.

Set heating power

Syntax: **setHP** *num* = 0

This command deactivates the PID regulator for the given cell and sets the heating power to a constant value. The heating power num is given as percentage and can be in the range of 0...100. If $num = 30$ then the heating power is set to 30%.

Fix heating power during PFC

Syntax: **fixHP** *oo* = off

When ΔT is set to a constant value the heating power is regulated by a PID controller. When this option is activated the PID controller is only active during the $1g$ phase. During all other phases the heating power is kept constant. The value of the heating power is calculated by a moving average over $25s$ of the values calculated by the PID controller. This option can help to counter PID controller related variations of the temperatures, which can have a negative influence on the calculation of $\mathcal{N}u$.

Pause the script

Syntax: **wait** *num* = 0

The script execution is paused for num seconds. All set parameters are kept at their current value and logging of the data continues. It can be used to collect data for a certain amount of time or to pause the script after setting a new parameter to e.g. allow the temperature to become stable before activating the electric field.

Activation of peltier elements

Syntax: **peltier** *cell oo* = off

This command activates (on) or deactivates (off) the peltier element connected to the given cell.

Activation of cooling pumps

Syntax: **pumpCool** *oo* = off

This command activates (on) or deactivates (off) the cooling pumps for all cells. Under normal operation of the experiment this pumps should always be activated before the peltier elements.

Activation of filling pumps

Syntax: **pumpFill** *oo* = off

This command activates (on) or deactivates (off) the filing pumps for all cells. During an experimental run this should not be necessary since the filling pumps are used as aid to fill the gap with the working fluid.

Activation of heating pumps

Syntax: **pumpHeat** *cell oo* = off

This command activates (on) or deactivates (off) the pump in the heating loop connected to the given cell. For normal operation these pumps should always be activated before setting the heating power or a temperature gradient.

Wait for microgravity

Syntax: **waitmg**

The execution of the script is paused until microgravity is reached ($g_z < 0.15g$).

Wait for completion of the next parabola

Syntax: **waitPara** *num* = 0

The execution of the script is paused until the next parabola is completed. The parabola is completed if, after a microgravity phase was reached, $0.9g < g_z < 1.1g$ is true for 10 seconds. If $num \neq 0$ an additional delay of num seconds is added.

Set marker for experimental runs

Syntax: **m** *num* = 0

This command sets the numerical marker to the given number. This can be used to distinguish between different parameter settings during the same experimental run. During a parabolic flight this number is set to correspond to the parabola number. Also the next and previous functions are based on this marker.

Activate data logging

Syntax: **log** *oo* = off

With this command the data logging can be activated or deactivated. When the process is activated the log file will be assigned a number starting from 0. When it is stopped this number is increased by 1, so that the following log section is written to a new file. Between the deactivation and reactivation of the log file should be a short delay of at least 1 second (*wait 1*).

Toggle cameras

Syntax: **camera** *cell oo* = off

With this command the camera assigned to the specified cell can be activated or deactivated.

Ramp the voltage over time

Syntax: **rampVoltage** $num_1 = 0$ $num_2 = 0$ $num_3 = 0$

With this function it is possible to ramp the voltage from num_1 to num_2 V_{peak} over num_3 minutes. Upon calling this command the voltage output is enabled and the peak voltage is set to num_1. After the duration of num_3 minutes has passed the output is disabled and the voltage is set to 0.

Activate the laser

Syntax: **laser** *oo* = off

This command activates or deactivates the laser. The activation of the laser is controlled by 3 independent mechanisms. First the Experiment Box needs to be closed, second the key switch needs to be activated, third the laser has to be activated by the LabVIEW program. This is required to comply with the safety requirements of operating laser systems.

Change high-voltage operation mode

Syntax: **laser** *str* = manual

There are currently two different operation modes for the high-voltage. In the default *manual* mode, the voltage is (de-)activated using the **voltage** command. In the *mg* mode, the voltage is only active when $g_z < 0.1g$.

Activate the mixing pumps

Syntax: **mix** *cell str* = off

The mix command controls the mixing pumps of the fluid loop in the gap. The *str* argument can be *on1* to pump in forward direction, *on2* to pump in reverse direction or *off* to deactivate the pumps.

C. Peltiercontroller

To improve the range of possible temperature gradients of the experiment, I developed a controller for the Peltier elements. It takes temperature readings of the cooling fluid using a PT-100 sensor. The Peltier element is controlled by a push-pull stage, which makes it possible to use the Peltier element for cooling and heating. The controller is connected to the PC via USB and can be controlled with LabVIEW. It has two input and output channels, so that two fluid loops can be controlled individually.

The temperature can either be acquired by a PT100 sensore, which is connected directly to the Peltier controller or it can be supplied by the communication interface. This was added to set the target temperature for the controller to the inlet of the cell, instead of directly at the heat exchanger. This way the losses from the heat exchanger to the cell are countered by the controller.

C.1. Temperature input

The signal from the PT100 is conditioned to a variable measurement range before it is read by the microcontroller. The result of the SPICE simulation is shown in figure C.1 and the final schematic in figure C.2.

The general way the circuit works is given by figure C.1. A constant current source with $1mA$ is used to power the PT100. The voltage drop over the sensor decoupled by a voltage follower ($U1$). At a temperature of $25°C$ this voltage is $109mV$. At $U2$ the reference voltage for the zero point is set. The lower end of the measurement range is set to $0°C$, which is equivalent to $100mV$ in the given setup. The reference voltage is subtracted from the input signal at the differential amplifier $U3$. The simulation shows the corrected signal as $13mV$. It would be expected to be $9mV$, but the input and output of the operational amplifier adds a offset to the signal. Finally the signal is amplified by 100 at the non-inverting amplifier stage at $U4$. With this values the measurement range for the temperatures is about $0°C$ to $100°C$. However, due to the used operational amplifiers in the simulation, the output signal is not linear anymore above $90°C$. For over-voltage and ESD protection a $3.6V$ Zener diode and TVS diode $TPD2E001$ are added.

In the final build of the circuit the $LM324$ (Figure C.2) are replaced by operational amplifiers $AD8608$, which offer Rail-to-Rain input and output. The constant current source based on the $LM317$ has been replaced. It is now based on a $LM334$ and $1N4148$, which are physically connected. This has been done to decrease the temperature drift of the current source, which is at $1.25\mu A/K$.

The final schematic (figure C.2) also includes several protection diodes and a analog low-pass filter at the output. The diodes protect against high voltages at the temperature

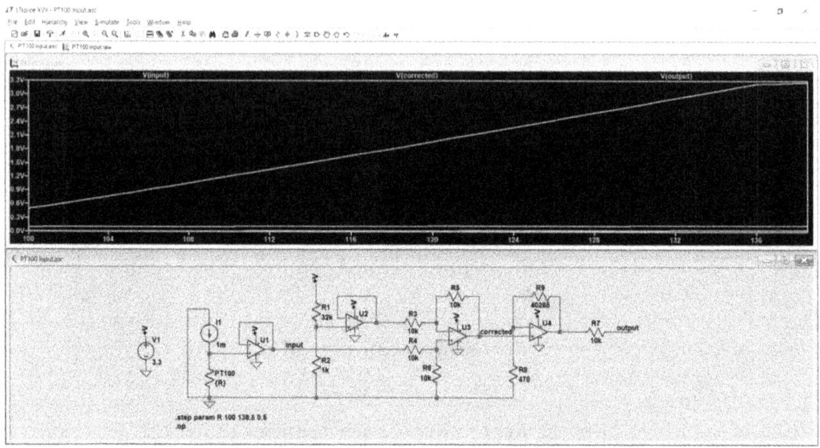

Figure C.1.: The SPICE simulation for the signal conditioning of the PT100 thermoelements. The curve shows a sweep of the PT100 from 100Ω to 138.5Ω on the x axis. This resistance range corresponds to a temperature range of $0°C$ to $100°C$. The y axis shows the response of the system. $V(input)$ is the input voltage after the voltage follower, $V(corrected)$ is the voltage after the differential amplifier and $V(output)$ is the output of the non-inverting amplifier.

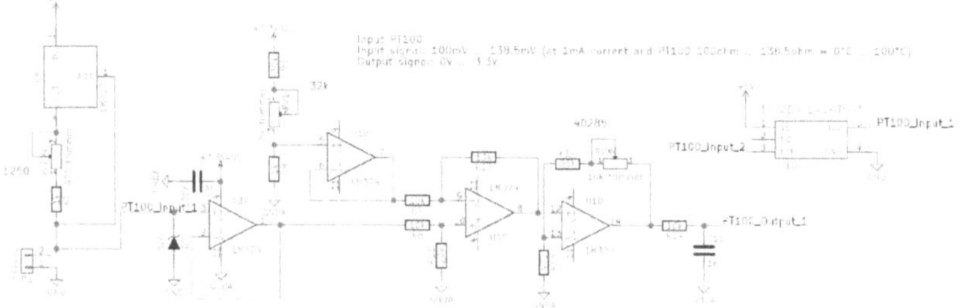

Figure C.2.: The schematic of the PT100 signal conditioning stages. In addition to the parts in the simulation Zener diodes ($3V6$) and TVS diodes ($TPD2E001$) are added to protect the operational amplifiers from voltage spikes.

input and surge currents, which may be cause by electrostatic discharges. The resistors responsible for setting the reference voltages are replaced by trimmer potentiometers. They can be used to offset for deviations in the parameters caused by production. These need to be adjusted after assembly of the system. To calibrate the values reference resistors have been used. When the PT100 is replaced by a 100Ω resistor the output of the system should be $0V$. The reference for the upper limit of the system is 120.5Ω, which is about $53°C$. This has been done to improve the resolution. When this reference is used, the output should be about $3.2V \dots 3.3V$. In a second calibration procedure, the temperature has been calibrated with an external reference as described in chapter 6.1.

C.2. Controller output

The push-pull stage, which controls the Peltier element are shown in figure C.3 and C.4. The simulation (figure C.3) shows the relation of input to output signal. The input signal in the range of $0V \dots 3.3V$ is compared to a reference set by $R1$ and $R2$. The difference amplifier $U1$ compares these voltages and translates them to a range of $-12V \dots + 12V$. The voltage follower $U3$ adjusts the actual output voltage based on the losses in the output stage. The resistor $R7$ is a replacement for the Peltier element.

In the actual schematics (figure C.4), the resistors to set the reference have been replaced by a trimmer potentiometer. This has to be adjusted in a way, that the output is $0V$, when the control signal is at half the maximum voltage (nominal $1.65V$). The gain at the difference amplifier can also be adjusted with a trimmer potentiometer. The output has to be $-12V$, when the input voltage is $3.3V$, and $+12V$, when the input voltage is $0V$.

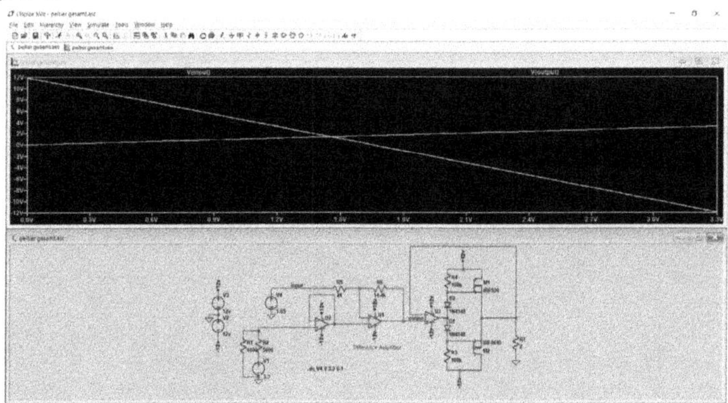

Figure C.3.: This figure shows the SPICE simulation of the output stage. An analog signal in the range of $0V \dots 3.3V$ is transformed to a signal of $-12V \dots + 12V$. The graph shows a voltage sweep from $0V$ to $3.3V$ of the control signal $V(input)$. $V(output)$ shows the output voltage to the driver stage.

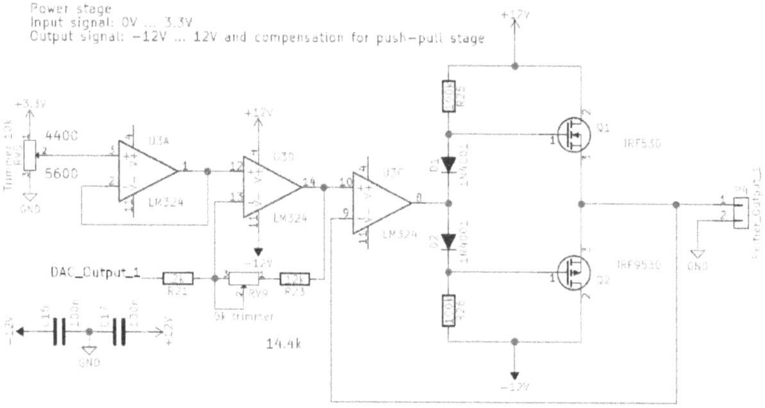

Figure C.4.: The schematic of the actual power stage. The reference voltage of the input signal at which the output is zero can be set at the trimmer potentiometer.

C.3. Data processing

The signals are processed by a STM32F072RB microcontroller. It has been chosen, because it has a 12-bit ADC, 12-bit DAC, and is also available as cheap development board (NUCLEO-F072RB), which saves a lot of time in developing this device. The controller uses the existing USB interface of the development board as virtual serial port to communicate with the PC.

The controller performs 10 temperature readings per second. Then a digital $2Hz$ low-pass filter is applied over the last 30 values to acquire the current actual temperature. This means that the system reacts slow to changes in the temperature, but is also very stable over time and has low deviations.

The regulation is based on a PID controller. It reads the current temperature once per second and adapts the output via the DAC of the controller. It is possible to set a threshold temperature. When this is reached, the controller will disable the output until the fluid has cooled down. There is a hysteresis to make sure the system does not oscillate. This kind of software shutdown is a safety requirement.

Before the Peltier controller can be used it needs to be calibrated. For calibration and debugging reasons I developed a specialized LabVIEW program. The first step is to set the lower temperature limit, which is $0°C$. The PT100 is put into a thermostat, filled with silicone oil AK5, and the temperature is set to $0°C$. In addition the temperature inside the thermostat was measured by a Checktemp 1 thermometer. When the temperature is stable, the temperature measured by the reference thermometer and the ADC values are written down. This is repeated with the upper temperature limit, which is $50°C$. All values, the temperatures and associated ADC values are written into the configuration of the controller. After that an additional calibration is done as described in chapter 6.1.

After the calibration process is done, the system was tested in the actual experimental setup. The target temperature for the fluid loop has been set to $22°C$. The temperature on both channels was recorded over a period of 20 minutes. The evaluation of this data gives an actual temperature of $22°C \pm 0.0143°C$ for channel 1 and $22°C \pm 0.0122°C$ for channel 2. The peak deviation from the set temperature was $0.05K$ for both channels. Heating the fluid loop from $20°C$ to $30°C$ takes about 11 minutes. This parameters are within the requirements set by the time frames of the parabolic flight procedure.

C.4. Communication protocol

The communication is realized with the virtual serial port, which is already implemented on the used development board. It uses a baudrate of 115200, 1 stop bit, 8 bit word length and no parity. The data is sent in packet. There are command packets, which the LabVIEW program sends to the controller, and response packets, which the controller sends back to confirm the execution of the requested command.

For both packets, the sequence of *0xFFA5* is used as synchronization. The packet length is then length of the whole packet, including synchronization and control bytes.

Table C.1.: The structure of the command packet, which is send from LabVIEW to the controller

Byte	0	1	2	3	4	...	n
Content	$0xFF$	$0xA5$	Packet length	Packet number	Command	Payload	Checksum

Table C.2.: The structure of the response packet, which is send from the controller to LabVIEW

Byte	0	1	2	3	4	...	n
Content	$0xFF$	$0xA5$	Packet length	Packet number	Status	Payload	Checksum

Since only one byte is used, the maximum packet length is 255 bytes and the maximum payload size is 249 bytes. The packet number is a ongoing number to distinguish between packets. The possible commands are explained in the next section in detail (Table C.6). The status byte in the response packet shows, if the requested command was executed successfully or not (Table C.3). The checksum is simply the sum of all bytes in the packet, excluding the checksum itself.

Table C.3.: Meanings of the possible status responses.

Code	Description
0	Success
1	Unknown command
2	Checksum incorrect
3	Other error

The data types used in the payload are standardized. Integers are transferred as big-endian with a length of x bytes, which is denoted as xb. Two types of decimal, fixed-point numbers are defined. The first type is $FP2$, which has 2 decimal places and is in the range of $-320.0 \ldots +320.0$. The second type $FP4$ has 4 decimal places and is in the range of $-2000.0 \ldots +2000.0$. If a command affects a given channel, it is denoted as $ch1$ for channel 1 and $ch2$ for channel 2.

In addition to these basic data types, the following packages are defined. The configuration of the channels is transferred in a config packet. The structure includes all data which are required for the temperature calibration and PID controller (Table C.4). The values of T_{min}, T_{max}, ADC_{min}, and ADC_{max} are used as first calibration, which defines a minimum and maximum temperature and linearizes the correlation over the measurement range. The coefficient and offset obtained from the calibration with an external reference are stored in $calib_c$ and $calib_o$. The k_p, k_i, and k_d values are the respective coefficients for the PID controller.

To access some of the internal values, which are required for the calibration, a debug packet is defined. It gives access to the current measured temperature, which is calcu-

Table C.4.: Structure of the config data.

Type	Name	Description
$FP2$	T_{min}	Lower end of temperature range
$FP2$	T_{max}	Upper end of temperature range
$2b$	ADC_{min}	ADC value ($uint16$) associated with lower temperature
$2b$	ADC_{max}	ADC value ($uint16$) associated with upper temperature
$FP4$	$calib_c$	Calibration coefficient
$FP4$	$calib_o$	Calibration offset
$FP4$	k_p	Proportional coefficient of PID controller
$FP4$	k_i	Integral coefficient of PID controller
$FP4$	k_d	Derivative coefficient of PID controller

lated using the values in the configuration. In addition the actual ADC value is shown, which makes it possible to see if the calculations are correct. The parameters e_{old}, e_{sum}, and PID_{out} allow to monitor the progression of the PID controller. This is required to make optimizations to the controller coefficients. The flags are used internally by the PID controller. The currently defined flags are $0x01$ to indicate whether the channel is active, $0x02$ to indicate whether the channel is running in constant power mode and $0x04$ to indicate that the channel is in overheated state.

Table C.5.: Structure of the debug data.

Type	Name	Description
$FP2$	$T_{measured}$	Measured temperature
$FP2$	T_{target}	Set target temperature
$2b$	ADC	Actual ADC value ($uint16$)
$2b$	DAC	Actual DAC value ($uint16$)
$FP4$	e_{old}	Previous error from PID controller
$FP4$	e_{sum}	Summed error from PID controller
$2b$	PID_{out}	Output from PID controller
$1b$	$flags$	Control flags

C.5. Explanation of the commands

$0x00$ - Synchronization

The command is used during the synchronization procedure. It is an empty packet, which allows to synchronize the receiver to the synchronization bytes of the sender.

Table C.6.: Supported commands and required payloads

Command	Description	Payload of command	Payload of response
$0x00$	Synchronization		
$0x01$	Request device information		$1b$ $0x18$ $1b$ Protocol version
$0x10$	Request configuration		$28b$ Config $ch1$ $28b$ Config $ch2$
$0x20$	Set new configuration	$28b$ Config $ch1$ $28b$ Config $ch2$	
$0x40$	Get temperature	$1b$ Channel	$FP2$ Temperature
$0x41$	Request data	Optional $FP2$ Temperature $ch1$ $FP2$ Temperature $ch2$	$1b$ PID active? $FP2$ Temperature $ch1$ $FP2$ Temperature $ch2$ $FP2$ Set Temp. $ch1$ $FP2$ Set Temp. $ch2$ $FP2$ Pelt. Temp. $ch1$ $FP2$ Pelt. Temp. $ch2$
$0x42$	Request debug information		$19b$ Debug $ch1$ $19b$ Debug $ch2$
$0x50$	Activate Peltier	$1b$ Channel	
$0x51$	Deactivate Peltier	$1b$ Channel	
$0x52$	Set target temperature	$1b$ Channel $FP2$ Temperature or $1b$ $0x03$ $FP2$ Temperature $ch1$ $FP2$ Temperature $ch2$	
$0x53$	Set constant power in %	$1b$ Channel $FP2$ Power% or $1b$ $0x03$ $FP2$ Power% $ch1$ $FP2$ Power% $ch2$	

$0x01$ - **Request device information**

This command is used once after successful synchronization. The expected response $0x18$ is a magic number as device identification. The protocol version is 1. If the commands or payloads change in a later development stage, the version can be used to ensure compatibility with older programs.

$0x10$ - **Request configuration**

The controller sends the current settings of the configuration to LabVIEW.

$0x20$ - **Set new configuration**

A new configuration for both channels is send to the controller.

$0x40$ - **Get temperature**

This command requests the actual temperature for channel 1 or 2.

$0x41$ - **Request data**

The program has two operation modes, which are set during compiling of the program. If the temperature is read from the PT100 at the controller, then the payload from LabVIEW to the controller can be omitted. But if the temperature is supplied externally the payload is required. The response includes whether the channels are active or not. If bit 1 in this byte is set, then the first channel is active. The second bit indicates activity of channel 2. Also the actual read temperatures and target temperatures of the controller are send. The last set target temperatures are send when the channel is inactive, but they do not influence the output anymore.

$0x42$ - **Request debug information**

This command is used by the debug and test LabVIEW program to request not only the temperatures, but also additional internal data. This makes it possible to debug the program flow on the controller. The *ADC* value in the debug packet is also required during the calibration process.

$0x50$ - **Activate Peltier**

Activates the PID controlled output for channel 1 or 2.

$0x51$ - **Deactivate Peltier**

Disables the output of channel 1 or 2 and also resets the PID controller for the channel.

$0x52$ - **Set target temperature**

Sets the target temperature for channel 1 or 2. The state of the PID controller is not changes by this command. If the channel is set to 3, then the temperature for both channels can be changed in the same packet.

$0x53$ - **Set constant power**

Sets the output power for channel 1 or 2 to a constant value. If the PID controller is running it is reset and disabled. The power% has to be in the range $-100 \ldots + 100$. At -100 the output is set to maximum cooling power. At -25 it is set to 25% of the maximum cooling power. A 0 disables the output and values > 0 will set the output to heat the Peltier element. If the channel is set to 3, then the power for both channels can be changed in the same packet.

C.6. Program flow

This section shows some flowcharts to explain the main data flow of the subprograms of the controller.

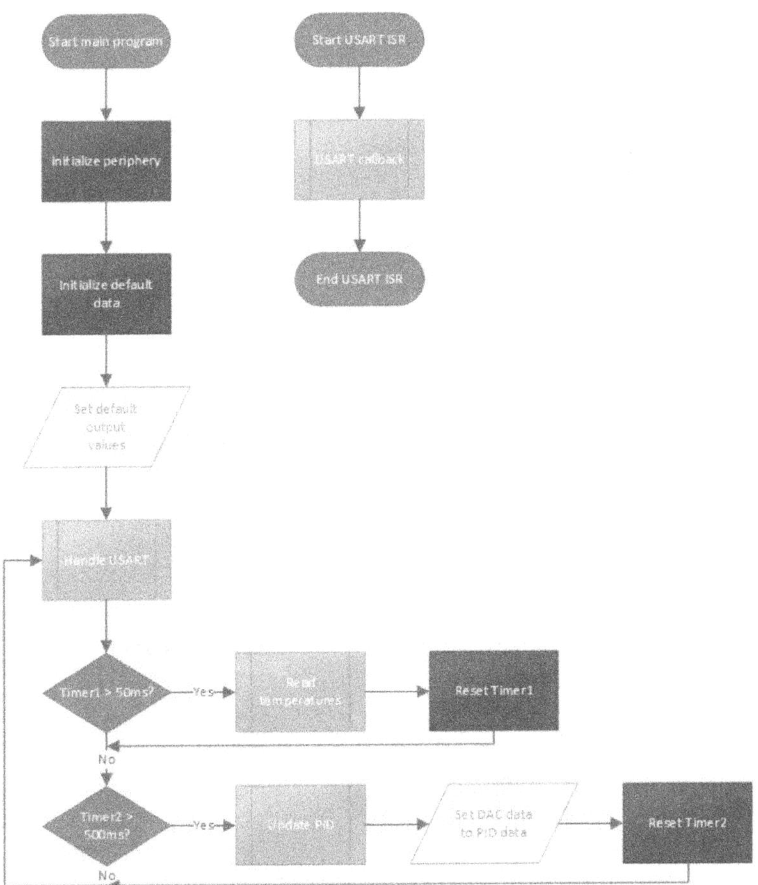

Figure C.5.: Flowchart of the main part of the program.

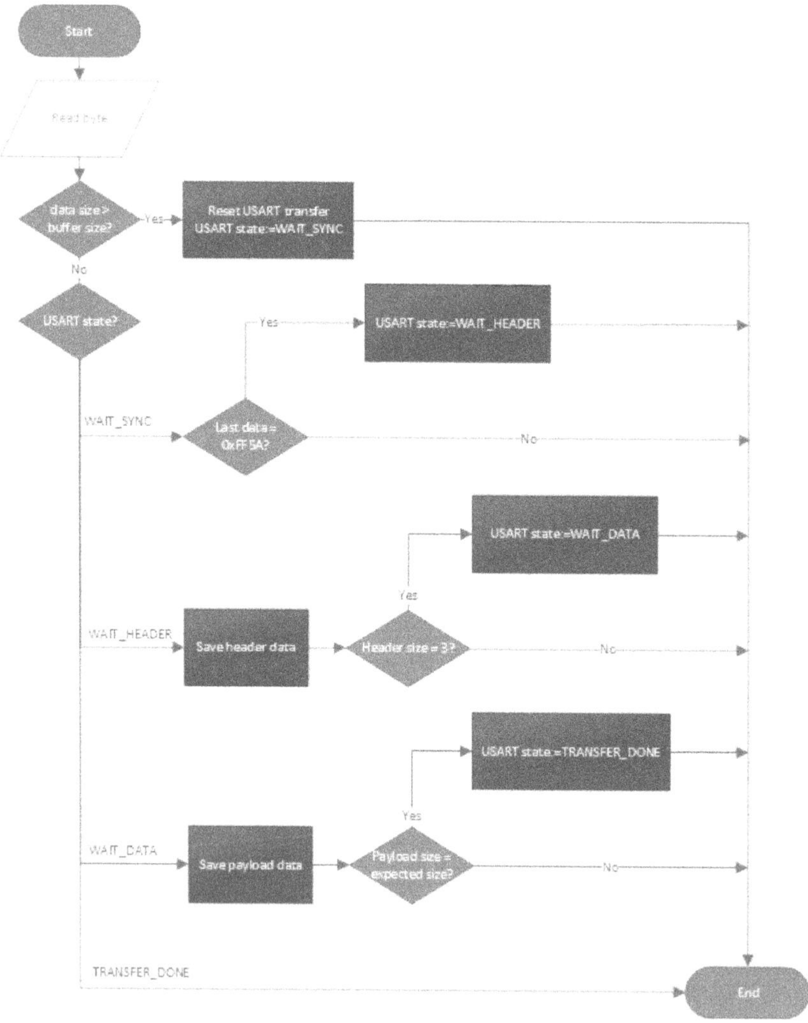

Figure C.6.: Flowchart of the interrupt service routine (ISR) which handles the data reception.

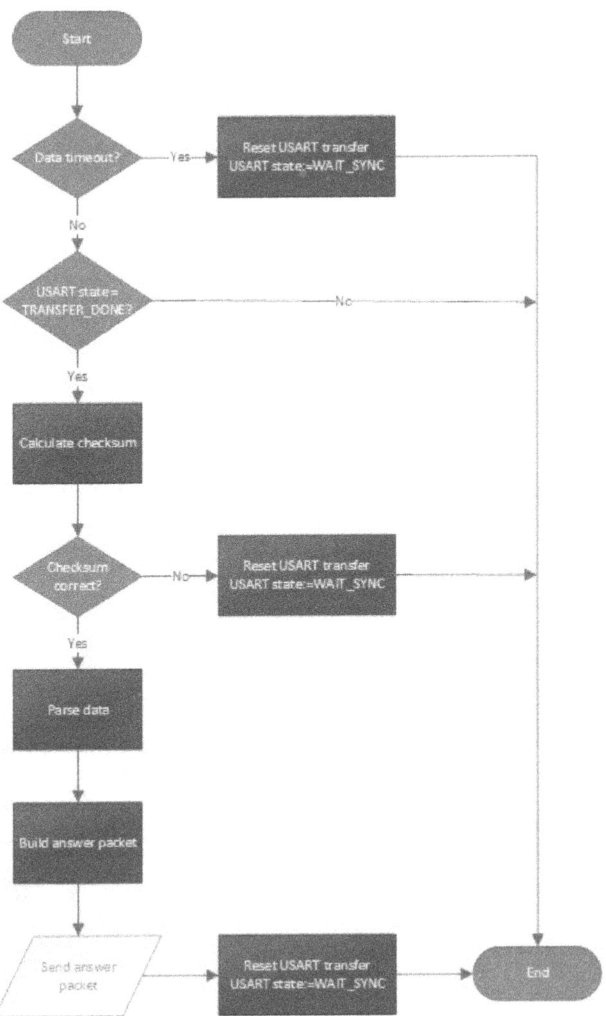

Figure C.7.: Flowchart of the USART handler which evaluates the received data.

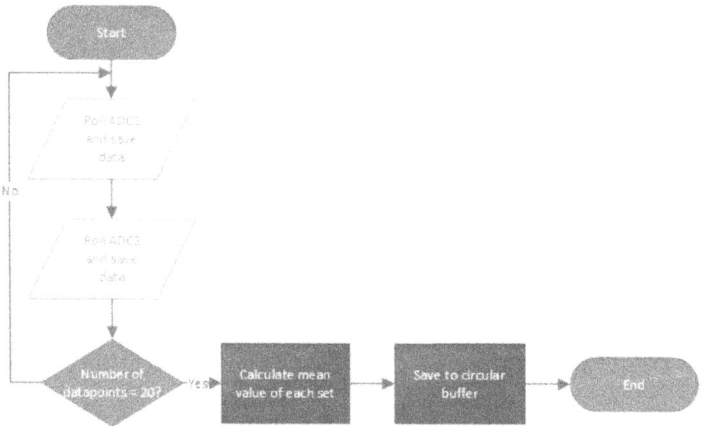

Figure C.8.: Flowchart of the subprogram which reads the analog values from the PT100.

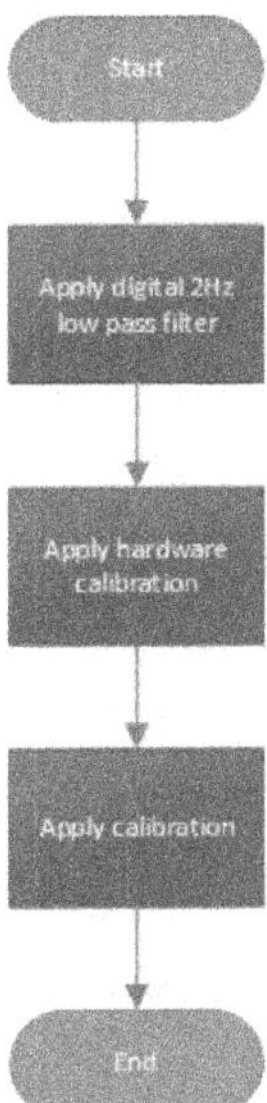

Figure C.9.: Flowchart of the subprogram which converts the analog values to an actual temperature. The hardware calibration is done by making reference measurements and setting the potentiometers on the hardware. Some of the resulting values are saved as part of the configuration. The second calibration is the same as for the other thermocouples used in the experimental setup (cf. chapter 6.1).

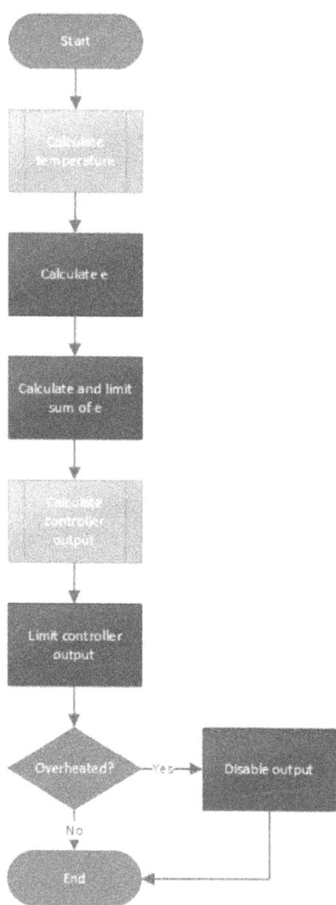

Figure C.10.: Flowchart of the subprogram which performs the PID control. The algorithm is comparable to the PID controller used in the LabVIEW program (cf. chapter B.2).

C.7. LabVIEW interface and debugging tool

The LabVIEW interface for debugging and calibrating the Peltier controller is designed very minimalist (Figure C.11). It allows to monitor not only the temperatures, but also the input at the ADC and output at the DAC. These data are required during the calibration process. The graphs can be used to track the development of the temperatures to see the reaction time of the controller and tune the PID coefficients accordingly.

The data flow is controlled based on the center control element. The COM port has to be set according to the number of the virtual COM port of the controller. The data in the configuration section have to be set according to the calibration.

The data transfer is package based. It is first waited until the identification sequence *0xFFA5* is received. Then the packet length is received and the program tries to receive this number of bytes. If the appropriate number of bytes is received, the checksum is checked and if this is correct, the packet is evaluated. If the required number of bytes is not received yet and no additional byte is received within $100ms$ of the last one, then a connection loss is assumed and the receive buffer is cleared and the transfer state is reset.

The system starts in a *disconnected* state. The COM port is opened using the LabVIEW API. A synchronization packet is send until a successful answer is received. If an unsuccessful answer is received, the data buffers of the connection is cleared and another synchronization is send. After synchronization, the device information is requested. On success, when the magic number and protocol version are correct, the state is changed to *connected* and the configuration data are transferred. A debug package is requested every $1s$. If no answer is received after $3s$, the LabVIEW program assumes that the controller has been disconnected and goes into a *disconnected* state and clears all buffered data.

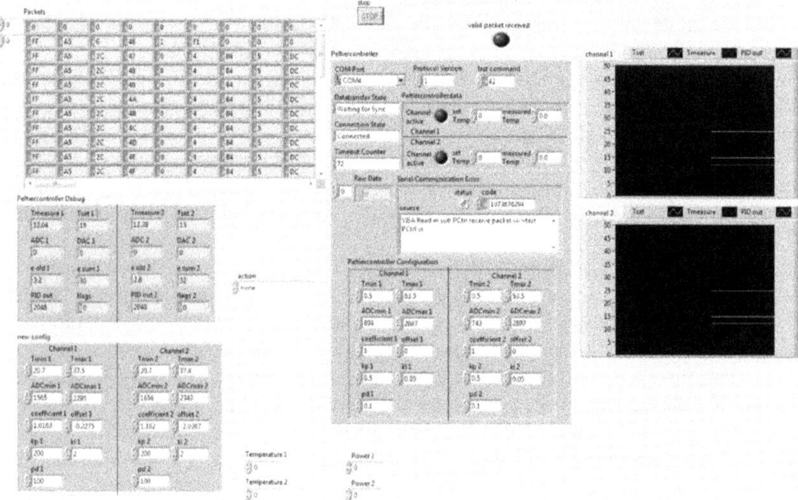

Figure C.11.: The debug interface for the Peltier controller in LabVIEW. The top left shows the listing of the received data packages as hexadecimal number. The middle left control element shows the debug data send by the controller. The lower left section allows to change the configuration of the controller. The center shows the main control element, which is also used in the main LabVIEW program. It contains all necessary data to control the data flow. The graphs on the right side are a graphical feedback of the current temperature, target temperature and heating power, so that the behavior of the controller can be analyzed.

www.ingramcontent.com/pod-product-compliance
Lightning Source LLC
Chambersburg PA
CBHW060309220326
41598CB00027B/4286

* 9 7 8 3 7 3 6 9 7 0 4 4 1 *